编著　新东方国外考试推广管理中心
上海新东方学校国外考试部

数学一本通

SAT

编委会

杨　鹏	钟　勋	张凯华	姜妍文	徐　晨
赵　鑫	田彦梅	陆凯伦	申振晖	张国一
桑　圆	刘烁炀	姚宇西	陈慧琳

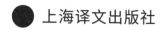 上海译文出版社

图书在版编目(CIP)数据

SAT数学一本通/上海新东方学校国外考试部等编著.
—上海：上海译文出版社，2020.1
ISBN 978 – 7 – 5327 – 8341 – 0

Ⅰ.① S… Ⅱ.① 上… Ⅲ.① 数学—高等学校—入学
考试—美国—教学参考资料 Ⅳ.① 01

中国版本图书馆 CIP 数据核字（2019）第 283863 号

SAT数学一本通
上海新东方学校国外考试部 等 编著
责任编辑/陆亚平 装帧设计/ideaforma工作室

上海译文出版社有限公司出版、发行
网址：www.yiwen.com.cn
200001 上海福建中路193号
上海市印刷四厂有限公司印刷

开本787×1092 1/16 印张10.75 字数207,000
2020年2月第1版 2020年2月第1次印刷
印数：0,001–5,000册

ISBN 978–7–5327–8341–0/H·1475
定价：42.00元

前　言

自 2016 年新 SAT 考试改革以来，数学史无前例地占据了总分的一半，对学生的最终得分起到了关键性作用。但与之脱节的是市场上数学教材的缺乏，学生对该科目的不重视，以及教育机构缺乏系统性的教学。目前官方教材只有《Official Guide》，其练习题也未进行知识点分类；巴郎或可汗学院练习题则过于陈旧，难易度不均，无法体现真实考试趋势。国内学生大体对数学过分自信，认为有了学校课堂习得的知识就可以拿高分，殊不知近年来数学分数水涨船高，考 750 分早已不具备竞争力。极少数 SAT 教育机构设立数学科目，但仍缺乏科学性的教学和练习。对于以美国名校为目标的学生来说，把握数学分数至关重要，全卷错 2 题以内，达到 780–800 的分数才算进入了优势区。

《SAT 数学一本通》的问世，不仅填补了市场上 SAT 数学教材短缺的空白，还为考生和教师提供了分级别的知识点练习，让备考或备课更有的放矢。《SAT 数学一本通》涵盖了新 SAT《Official Guide》数学考纲中的所有考点，共分 24 个章节：第 1–8 章为代数部分，第 9–15 章为解决问题和数据分析能力部分，第 16–17 章为高等数学部分，第 18–24 章为其他部分。所有题目都与 2016–2018 年真题相仿，并配以难度分级：LEVEL 1 适合 700 分以下考生；LEVEL 2 适合 700–750 分考生；LEVEL 3 适合 750 分以上考生。未参加过实考的学生建议结合 SAT 数学课程，循序渐进地完成全书的练习；已实考同学建议结合成绩单上的小分数（即 subscore）提示自行分级练习。无论何种基础的学生都能从中找到与自身能力匹配的练习，更有针对性地备考。本书的最后编入了两套模拟题，供考生在系统学习完知识点后自我检测，尽快进入临场考试状态。

本书的编写工作能顺利完成，得益于上海新东方积累十年的考试秘笈，更体现了数学教研组强大的集体智慧。编委会成员中不乏具有多年教学经验的骨干教师，还汇聚了大量世界名牌大学理学或工学硕士及博士。教研组每周召开课程研讨会，定期修订并优化教材内容。

感谢新东方集团领导和上海新东方项目主管张凯华老师为此书提供了丰富的资源，感谢全体编委会成员为此书付出的辛勤汗水和宝贵意见。

最后，我们衷心祝愿广大考生取得理想成绩！

姜妍文

2020 年元月

于上海新东方学校

目 录

1. Equation

LEVEL 1

1

$$\frac{3}{x+2} + \frac{5}{3(x+2)}$$

Which of the following expressions is equivalent to the one above, where $x \neq -2$?

A) $\frac{14}{x+2}$

B) $\frac{14}{3x+6}$

C) $\frac{4}{3x+6}$

D) $\frac{-4}{x+2}$

2

$$2x^2 + 2y^2 = 40$$

$$y - \frac{x}{2} = 0$$

If the ordered pair (x, y) is a solution to the system of equations above, and $x>0$, what is the value of y?

A) -2

B) 2

C) 4

D) -8

3

If $m = 3(x-2)^2$ and $n = 4(x-2)^3$, which of the following is equivalent to $\frac{m}{n}$?

A) $\frac{3}{4}(x-2)^5$

B) $\frac{3}{4}(x-2)$

C) $\frac{3}{4(x-2)}$

D) $\frac{3}{4(x-2)^5}$

4

A clothing store sells T-shirts for $10 each and shorts for $20 each. A customer spends $150 on shorts and T-shirts. If the customer buys 3 fewer T-shirts than shorts, how many T-shirts did the customer buy?

A) 3

B) 4

C) 5

D) 6

5

If $3(a + 2) = 4(b - 3)$, what is a in terms of b?

A) $a = \dfrac{4b - 6}{3}$

B) $a = \dfrac{4b - 12}{3}$

C) $a = \dfrac{4b - 18}{3}$

D) $a = \dfrac{4b - 18}{4}$

6

If 5 times m equal to 20, what is the value of $m + 8$?

A) 10

B) 12

C) 16

D) 9

7

If $\dfrac{1}{2}cx - 5 = 12$, what is the value of $cx - 5$?

A) 20

B) 27

C) 29

D) 9

8

$$x(x + 1) = 12$$

Which of the following lists are solutions to the quadratic equation above?

A) 0 and –2

B) –4 and 3

C) 4 and 3

D) $\sqrt{8}$

9

$$1.2p = t$$

At a liquor store, a wine originally priced at p dollars and now it increased to t dollars, and the relationship between p and t is given in the equation above. What is p in terms of t?

A) $p = t + 1.2$

B) $p = 1.2t$

C) $p = \dfrac{1.2}{t}$

D) $p = \dfrac{t}{1.2}$

10

If $b = 5$, how much greater is $10b$ than $4b$?

A) 20

B) 30

C) 40

D) 50

11

Which of the following is equivalent to the expression $7(x-1)^2 - 6(x-1)^2$?

A) 1

B) $x^2 - 1$

C) $x^2 - 2x + 1$

D) $x^2 - 2x + 13$

12

$$c^2 - 6c + 9 = 25$$

Based in the equation above, which of the following could be the value of $c - 3$?

A) 5

B) 4

C) 2

D) 1

13

$$7x^2 - 5(2 - x) + x(2 + x)$$

Which of the following polynomial is equivalent to the expression above?

A) $8x^2 + 7x - 10$

B) $8x^2 - 7x - 10$

C) $8x^2 - 7x + 10$

D) $7x^2 + 8x - 10$

14

If $-2x + 5 = -x - 4$, what is the value of x?

A) -9

B) 9

C) -1

D) 1

15

If $\dfrac{30}{2x - 5} = 4$, what is the value of $x + 2$?

A) 7.25

B) 7.5

C) 8.25

D) 8.5

16

$$\frac{3}{s+2} - \frac{7}{5(s+2)}$$

Which of the following expressions is equivalent to the one above, where $s \neq -2$?

A) $\dfrac{8}{5(s+2)}$

B) $\dfrac{11}{5(s+2)}$

C) $\dfrac{8}{s+2}$

D) $\dfrac{10}{s+2}$

17

If $m = -2$, how much smaller is $7m$ than $2m$?

A) 15

B) –15

C) 10

D) –10

18

During a rainy day, the precipitation of Shanghai keeps at a constant 28 mm per hour over a 24-hour time period. Which of the following is closest to the total precipitation in mm, in 7 hours?

A) 230

B) 196

C) 198

D) 197

19

If $x \neq 3$, what is the value of

$\dfrac{5}{3x - 9} \times (2x - 6)?$

20

$$5(x + y) = x$$

If (x, y) is a solution to the equation above and $x \neq 0$, what is the ratio $\dfrac{y}{x}$?

A) $-\dfrac{6}{5}$

B) $-\dfrac{4}{5}$

C) $\dfrac{1}{5}$

D) $\dfrac{2}{5}$

21

What is the value of t if

$(t + 2) - (4t - 7) = -3?$

22

For what value of m is $12 = \dfrac{m}{3} - 8?$

A sock store sells a brand of socks in individual pairs and in packs of 5 pairs. On a certain day, the store sold a total of 312 pairs of the brand of socks, of which 32 were sold as individual pairs. Which equation shows the number of packs of pairs, ρ, sold that day?

A) $\rho = \dfrac{312 - 32}{5}$

B) $\rho = \dfrac{312 + 32}{5}$

C) $\rho = \dfrac{312}{5} - 32$

D) $\rho = \dfrac{312}{5} + 32$

Which ordered pair (x, y) satisfies the system of equations shown below?

$$2x - y = 12$$
$$x + 2y = -4$$

A) $(-12, 4)$
B) $(-4, 4)$
C) $(4, -4)$
D) $(8, 4)$

A truck traveled at an average speed of 70 miles per hour for 4 hours and consumed fuel at a rate of 41 miles per gallon. Approximately how many gallons of fuel did the car use for the entire 4-hour trip?

A) 2
B) 3
C) 5
D) 7

If $\dfrac{7}{x} = 140$, what is the value of x?

A) 980
B) 70
C) 20
D) 0.05

What value of x satisfies the equation $5x + 4 = 24$?

A) 2
B) 4
C) 6
D) 20

If $\frac{3n}{7} = 14$, what is the value of $6n - 1$?

A) 2

B) 3

C) 97

D) 195

$$3(x + 5) = 7 + 5x$$

What value of x satisfies the equation above?

$$3(2 - x) = 6x + 13$$

Which of the following is the solution to the equation above?

A) $-\frac{7}{3}$

B) $\frac{19}{3}$

C) $\frac{7}{9}$

D) $-\frac{7}{9}$

$$5x + y = 2$$
$$y = 4x - 7$$

In the solution to the system of equations above, what is the value of x?

A) 1

B) $\frac{5}{9}$

C) 9

D) -5

$$3x + 9 = 6$$

If x satisfies the equation above, what is the value of $x + 3$?

LEVEL 2

1

If $\frac{2}{5}x + 3 = k + \frac{1}{4}x$, what is the value of k when $x = 0$?

A) -3

B) 0

C) 1

D) 3

2

Note: figure not drawn to scale

In the figure above, AC = 27. If $y = 4x$, what is the length of the line BD?

A) 4.5

B) 18

C) 31.5

D) 41.5

3

A toy company sold 50 toys that consist of t toy trains and c toy cars. The company sold the trains for $2 each and sold the cars $3 each and collected a total of $100. Which of the following systems of equations can be used to find the number of cars sold?

A) $t + c = 100$
   $2t + 3c = 50$

B) $t + c = 50$
   $2t + 3c = 100$

C) $t + c = 100$
   $3t + 2c = 50$

D) $t + c = 50$
   $3t + 2c = 100$

4

$$x - 3y = 10$$
$$2x + 6y = 20$$

The solution to the system of equations above is (x, y). What is the value of x?

A) 0

B) -10

C) 10

D) 20

5

At a coffee house, each Americano costs $4 and each Latte costs d more dollars than an Americano. If 3 Americano and 4 Latte cost a total of $30. What is the value of d?

A) 0.5

B) 1

C) 1.2

D) 2

6

When $3x$ is multiplied by 5 and then increased by 3, the result is less than 48. What is the largest possible integer value for x?

7

Accommodation fee by distance from Downtown

Distance from downtown	Accommodation fee
30-50 miles	$50
10-29 miles	$100
1-9 miles	$500

Accommodation fee is based on the distance of hotel from downtown, as shown in the table above. A group of 8 people want to get 8 rooms: 2 people will book hotel 45 miles away from downtown; 3 people will book hotel 12 miles away from downtown; 3 people will book hotel 2 miles away from downtown. What is the average (arithmetic mean) accommodation fee, in dollars, for 8 rooms? (Disregard the $ sign when gridding your answer.)

8

$$y = x^2 - 2x$$
$$x = y + 2$$

The system of equations above is graphed in the xy-plane. Which of the following is the y-coordinate of an intersection point (x, y) of the graphs of the two equations?

A) 1

B) –1

C) 2

D) –2

9

If $4a^2b^2 = 9$ and $ab > 0$, what is the value of $10ab$?

A) 10

B) 15

C) 50

D) 100

$$3x + 4y = 5$$
$$x + y = 3$$

What is the value of x in the system of linear equations above?

If x is positive and $(x - 4) + (x + 1) + x = 5(x^2 - 1)$, what is the value of x?

$$3x + 4y = 20$$
$$\frac{1}{2}x + y = \frac{9}{2}$$

If (x, y) satisfies the system of equations above, what is the value of y?

$x^6 - y^6 = 35$ and $x^3 + y^3 = 5$, what is the value of $x^3 - y^3$?

A) 4

B) 5

C) 6

D) 7

Which of the following is equivalent to the expression $\frac{2m + 6}{4} \times \frac{6m - 36}{3m + 9}$?

A) $\frac{12m^2 - 216}{12m + 36}$

B) $\frac{8m - 30}{3m + 13}$

C) $\frac{m - 6}{4}$

D) $m - 6$

$$2c + 3d = 17$$
$$6c + 5d = 39$$

Based on the system of equations above, what is the value of $4c + 4d$?

A) –4

B) 1

C) 28

D) 56

16

Which of the following is equivalent to the product of the expression $(2x - 3)(x + 1)$?

A) $2x^2 - x - 3$

B) $2x^2 - x + 3$

C) $2x^2 + x - 3$

D) $2x^2 + x + 3$

17

Michael plants 765 flowers. Some of them are lilies with 6 petals, and some are roses with 9 petals. The sum of the petals is 6,009. How many more roses than lilies did Michael planted?

18

Andy mixed milk candies with strawberry candies to make a candy package for sale. Each package contains 1.1 pounds of milk candies and 2.3 pounds of strawberry candies and the total cost of the candies is 62 dollars. Each pound of milk candies cost 12 dollars. How much does each pound of strawberry candy cost, to the nearest dollar?

19

$$x - \frac{1}{2}y = 10$$

$$\frac{1}{4}x - \frac{1}{4}y = 19$$

Which ordered pair (x, y) satisfies the system of equations above?

A) $(-56, -132)$

B) $(32, 44)$

C) $(\frac{116}{3}, \frac{112}{3})$

D) $(144, 268)$

20

Students are having lunch in the school. Each boy needs 3 pieces of bread and each girl needs 2 pieces of bread. There are 13 children in total having lunch and the kindergarten prepared 33 pieces of bread to meet everyone's need. How many boys are there having lunch?

21

Grace wants to buy skirts that cost $37.30 each and coats that cost $45.80 each. A 6% sales tax will be applied to the entire purchase. If Grace buys 2 skirts, which equation relates the number of coats purchased, c, and the total cost in dollars, t?

A) $1.06(74.60 + 45.80c) = t$
B) $74.60 + 45.80c = 0.94t$
C) $74.60 + 45.80c = 1.06t$
D) $0.94(74.60 + 45.80c) = t$

22

$$3x + 9 = 6$$

If x satisfies the equation above, what is the value of $6x + 18$?

23

If $2w + 4t = 12$ and $4w - 6t = 3$, what is the result of $w + 2t$?

A) 2
B) 0
C) 4
D) 6

24

First-time customers at a cat-boarding facility must pay a onetime $20 evaluation fee, as well as $15 for each day and $10 for each night the cat spends at the facility. Which of the following expressions best models the total cost, in dollars, for a first-time customer to board a cat at this facility for d days and n nights?

A) $15d + 10n + 20$
B) $(20 + 15)d + 10n$
C) $(20 + 15)(d + n)$
D) $(20 + 15)(d + n) - 10$

25

$$y = x - 5$$
$$y = -3x + 31$$

If (a, b) is the point of intersection of the two lines in the xy-plane represented by the equation above, what is the value of a?

LEVEL 3

$$2x - 3y = 11$$
$$3x + 5y = 23$$

The solution to the system of equations above is (x, y), what is the value of y?

A) 12

B) $\frac{13}{19}$

C) $\frac{5}{6}$

D) $\frac{3}{19}$

If the expression of $2a^2b^2 - 7ab + 3$ is equivalent to $(2ab - m)(ab - n)$, where m and n are constants. What is one possible value of $m + n$?

A) 1

B) 2

C) 3

D) 4

Simon read his favorite book Harry Potter for three times as many minutes on Sunday as he did on Saturday. He read the Harry Potter those two days for a total of 3 hours and 20 minutes. For how many minutes did Simon read his favorite book on Sunday?

Questions 4–5 refer to the following information.

A fitness club is hosting an indoor triathlon. Each triathlon participant spends 10 minutes swimming in the pool, 20 minutes biking on a stationary bike, and 30 minutes running on a treadmill. Scores are given based on total combined distance in miles, D, of the three activities, which can be found using the equation below.

4

$$\frac{1}{6}s + \frac{1}{3}b + \frac{1}{2}r = D$$

In the equation, s is the participant's average swimming speed, in miles per hour (mph), b is the participant's average biking speed, in mph, and r is the participant's average running speed, in mph.

Kate's average running speed during the triathlon was 7.5 mph, and her average biking speed was 16 mph. Her total combined distance was 10.00 miles. At what average speed, in mph, did she swim?

A) 1.50

B) 3.50

C) 5.50

D) 7.50

5

In the triathlon, Jane ran exactly 6 times the distance she swam. Which equation relates her average swimming speed to her average running speed?

A) $r = 6s$

B) $r = 3s$

C) $r = 2s$

D) $r = \frac{1}{2}s$

2. Absolute Value

LEVEL 2

1

$$|8 + x| = 23$$

The value of one solution to the equation is 15, what is the value of the other solution?

A) 31

B) –31

C) 41

D) –41

2

$$|x - 7| = 3$$

The value of one solution to the equation above is 10. What is the value of the other solution?

3. Inequality

1

$$6x - 3y > 9$$

Which of the following ordered pairs (x, y) are in the solution set for the inequality above?

I. $(-1, -9)$ II. $(3, 0)$ III. $(1, 3)$

A) I only

B) II only

C) III only

D) I and II only

2

A store in a university sells a pencil for $2 each and an eraser for $1 each. If the total income for one day is $50, and at least 10 of the erasers were sold, what is the maximum number of pencils that could have been sold that day?

A) 19

B) 20

C) 21

D) 22

3

If $-3 < 2x-1 < -1$, what is one possible value of $2 - 4x$?

4

If $f(x) = x^2 - 2x + 5$ and b is a positive integer less than 4, what is one possible value of $f(b)$?

5

When 5 is decreased by $6x$, the result is less than -24, what is the smallest possible integer value for x?

6

A woman has \$20 to buy fruits. She wants to buy both bananas and pears. Bananas cost \$1.99 per kilogram, and pears cost \$0.99 per kilogram. If b represents the number of kilograms of bananas and p represents the number of kilograms of pears, which of the following systems of inequalities models this situation?

A) $\begin{cases} 1.99b + 0.99p \geq 20 \\ b + p \leq 1 \end{cases}$

B) $\begin{cases} 1.99b + 0.99p \leq 20 \\ b + p \leq 1 \end{cases}$

C) $\begin{cases} 1.99b + 0.99p \geq 20 \\ b > 0 \\ p > 0 \end{cases}$

D) $\begin{cases} 1.99b + 0.99p \leq 20 \\ b > 0 \\ p > 0 \end{cases}$

LEVEL 2

1

In an electronic store, headphones cost \$5 each and chargers cost \$10 each. If a group of customers spends at least \$50 but at most 80\$ on 5 headphones and n chargers, which of the following is true about n?

A) $12 \leq n \leq 70$

B) $10 \leq n \leq 12$

C) $8 \leq n \leq 10$

D) $3 \leq n \leq 5$

2

The table below shows the required age of members to take different classes in a gym. In which of the following inequalities does a represent the age range of the three types of weight lifting classes listed in the table?

Types of classes	Required age
Yoga	13-16
Weight lifting (starter)	17-18
Weight lifting (intermediate)	19-21
Weight lifting (expert)	22-25
Boxing	26-30

A) $a \leq 30$

B) $17 \leq a$

C) $17 \leq a \leq 25$

D) $17 \leq a \leq 18$

3

At a supermarket, apples cost \$2 each and bananas cost \$3 each. If 100 of them were sold, and the income is between \$220 and \$223, inclusive, what is one possible number of apples that were sold?

4

Shawn has \$2,000. He wants to buy at least 3 headphones and at least 4 laptops for his friends and colleagues. Each headphone costs \$200 and each laptop costs \$300. If h represents the number of headphones and l represents the number of laptops, which of the following systems of inequalities represents this situation?

A) $\begin{cases} h + l \leq 2,000 \\ h \geq 3 \\ l \geq 4 \end{cases}$

B) $\begin{cases} 3h + 4l \leq 2,000 \\ h \geq 3 \\ l \geq 4 \end{cases}$

C) $\begin{cases} 3h + 4l \leq 2,000 \\ h \geq 200 \\ l \geq 300 \end{cases}$

D) $\begin{cases} 200h + 300l \leq 2,000 \\ h \geq 3 \\ l \geq 4 \end{cases}$

5

$$s = 20x + 8y$$

The formula above gives the daily salary of a salesman, in dollars, of selling different clothes when the salesman sells x shirts and y scarves. If, in a particular day, the man earns no more than \$300 and at most 12 shirts were sold, which is the maximum number of scarves the salesman could have sold?

A) 7

B) 6

C) 8

D) 5

6

John Watson wants to buy some fruits for Sherlock and himself. He found that there were only £50.89 in his purse. The apple was sold for £5.31 per pound and he bought a pounds, and banana was £4.93 per pound and he bought b pounds. The total mass of fruits was no more than 9 pounds. Which of the following systems of inequalities can be used to find the number of pounds of apple and banana?

A) $\begin{cases} a + b = 9 \\ 5.31a + 4.93b = 50.89 \end{cases}$

B) $\begin{cases} a + b \leq 9 \\ 5.31a + 4.93b \leq 50.89 \end{cases}$

C) $\begin{cases} a + b \leq 50.89 \\ 5.31a + 4.93b \leq 9 \end{cases}$

D) $\begin{cases} a + b \leq 9 \\ 4.93a + 5.31b \leq 50.89 \end{cases}$

To clean a kitchen, Shawn charges a fee of $20 for his tools and $10.50 per hour spent cleaning a kitchen. Jeremy charges a fee of $17 for his tools and $11.25 per hour spent cleaning a kitchen. If h represents the number of hours spent cleaning a kitchen, what are all values of h for which Jeremy's total charge is greater than Shawn's total charge?

A) $h > 4$

B) $3 \leq h < 4$

C) $4 \leq h \leq 5$

D) $h < 3$

A juice company is filling bottles of juice from a tank that contains 200 gallons of juice. At most, how many 40-ounce bottles can be filled from the tank? (1 gallon = 128 ounces)

A) 5

B) 108

C) 128

D) 640

There are 12 apples in a box. Each apple is at least 200 grams and no more than 300 grams. Which inequality represents all possible value of the total weight of the apples, w, in grams?

A) $200 \leq w \leq 300$

B) $2,400 \leq w \leq 2,500$

C) $2,400 \leq w \leq 3,600$

D) $3,500 \leq w \leq 3,600$

A cargo helicopter delivers only 100-pound packages and 120-pound packages. For each delivery trip, the helicopter must carry at least 12 packages, and the total weight of the packages can be at most 1,400 pounds. What is the maximum number of 120-pound packages that the helicopter can carry per trip?

A) 2

B) 8

C) 10

D) 12

LEVEL 3

1

$$C = 5x + 2y$$

The formula above gives the monthly cost C, in dollars, of operating an Uber when the driver works a total of x hours and when y gallons of gasoline are used. If, in a particular month, it costs no more than $1,000 to operate the Uber and at least 200 gallons of gas were used, what is the maximum number of hours the driver could have worked?

A) 100

B) 120

C) 200

D) 1,400

2

A floral decoration is required for the opening of a new building. The decoration must be at most 30 meters long and is to be made up of two types of flowers: type A and type B. Type A is 1 meter in length, and Type B is 1.5 meter in length. The client wants at least twice as many as type A as type B, and at least 6 of type B. If a represents the number of type A and b represents the number of type B. Which of the following systems of inequalities represents this situation?

A) $\begin{cases} a + 1.5b \leq 30 \\ a \leq 2b \\ b \geq 6 \end{cases}$

B) $\begin{cases} a + 1.5b \leq 30 \\ a \geq 2b \\ b \geq 6 \end{cases}$

C) $\begin{cases} a + 1.5b \geq 30 \\ a \leq 2b \\ b \geq 6 \end{cases}$

D) $\begin{cases} a + 1.5b \geq 30 \\ a \geq 2b \\ b \geq 6 \end{cases}$

4. Linear Function

LEVEL 1

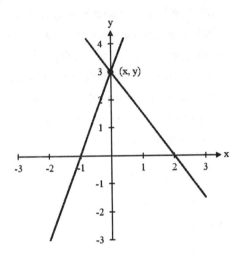

In the *xy*-plane, what is the solution (x, y) to the system of the equations graphed above?

A) $(0, 3)$

B) $(2, 0)$

C) $(1, 2)$

D) $(0, 1)$

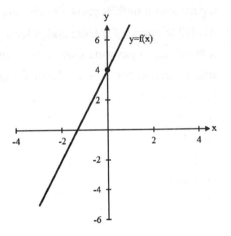

The line in the *xy*-plane is the graph of the linear function *f*, which has a slope of 3 and contains the point $(0, 4)$. What is the value of $f(3)$?

A) 9

B) 13

C) 12

D) 5

3

$$f = 20d + 100$$

The number of total funds, f, in Tom's company d days after they opened the store is modeled by equation above. Which of the following best describes the meaning of the number 100 in the equation?

A) The number of funds Tom invested into the store each day

B) The number of funds Tom initially invested into the store

C) The total number of funds in the store after d days

D) The number of days since the store was opened

4

Which of the following examples would exhibit linear growth over time?

A) A watch increasing in value by 3% every year

B) The number of population in a city decreasing by 20% every year

C) The distance of walks by an athlete doubling every day

D) A watch increasing in value by $50 every year

5

Eric opens a new coffee shop near Lindon High School. It costs $2 to make a coffee. If he sells each cup for $7, which of the following equations gives the amount of profit p, in dollars, Eric will make for selling n cups?

A) $p = 5n$

B) $p = 7n$

C) $p = 7n - 2$

D) $p = 7n + 2$

6

Richard is 150 pounds and is gaining weight at a constant rate of 0.5 pound per month. At this rate, in how many months will Richard be at least 157 pounds?

7

Emily is 1.5 m in height and is getting taller at a constant rate of 2 cm per year. Which of the following function best approximates the height $h(t)$, in meter, t years after, for $0 \leq t \leq 30$?

A) $h(t) = 1.5 + 0.02t$

B) $h(t) = 1.5 - 0.02t$

C) $h(t) = 1.5 + 0.3t$

D) $h(t) = 1.5 - 0.3t$

Which of the following is the graph of equation $y = 3x + 1$ in the xy-plane?

A)
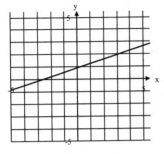

B)
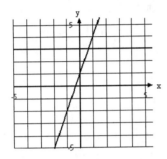

C)
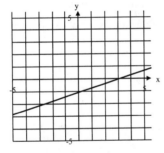

D)
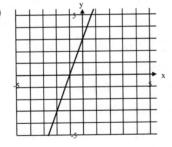

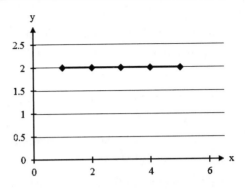

Which of the following is an equation of the line shown in the xy-plane above?

A) $x = 2$

B) $y = 2$

C) $x = y$

D) $x = 3y$

John took taxi to work. The fee f, in pounds, of the taxi is estimated by the equation $f(x) = 18 + 2.5x$, where x is the distance travelled in km. What is the best interpretation of the number 2.5?

A) The estimated fee, in pounds, of 1 km

B) The estimated fee, in pounds, of 18 km

C) The estimated fee, in pounds, of the taxi with no distance

D) The estimated fee, in pounds, of 2.5 km

11

In the *xy*-plane, what is the slope of the line that passes through the points (2, 3) and (–7, 11)

A) $\frac{8}{9}$

B) $-\frac{8}{9}$

C) $\frac{14}{5}$

D) $-\frac{14}{5}$

12

In the *xy*-plane, what is the *y*-intercept of the line that has a slope of 1.5 and passes the point (2, 9)?

A) 4

B) 5

C) 6

D) 7

13

$$R = 2000 + 50x$$

A bar uses the function above to estimate the revenue R, in dollars, when x customers came to the bar each week. Based on the model, how many customers may come if the revenue is 4,200 dollars?

14

$$h = -\frac{1}{175}t + 480$$

What is the value of t in terms of h?

A) $t = -175 \times (h - 480)$

B) $t = 175 \times (h - 480)$

C) $t = -175 \times (h + 480)$

D) $t = 175 \times (h + 480)$

Which of the following is the graph of the equation $y = -3x + 7$ in the xy-plane?

A)

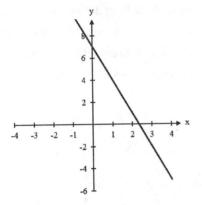

B)

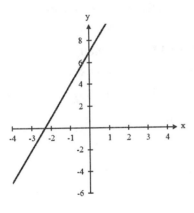

C)

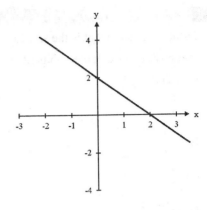

D)

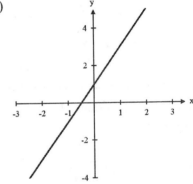

A cook earns $30 per hour for preparing 2 kinds of dishes for the restaurant. In addition, he can get a onetime tip for $10 if he cleans up the kitchen. If he cleans up the kitchen, what expression could be used to determine how much he will earn?

A) $30x + 10$, where x is the number of hours

B) $10x + 30$, where x is the number of hours

C) $x(30 + 2) + 10$, where x is the number of dishes

D) $10x + (30 + 2)$, where x is the number of dishes

$$n = 512 + 7T$$

The equation above is used to model the relationship between the number of ice creams, n, sold per day in an ice cream shop and the average daily temperature, T, in degrees Fahrenheit. According to the model, what is the meaning of the 7 in the equation?

A) For every increase of 7°F, one more ice cream will be sold.

B) For every decrease of 7°F, one more ice cream will be sold.

C) For every increase of 1°F, seven more ice creams will be sold.

D) For every decrease of 1°F, seven more ice creams will be sold.

Eva subscribes to a music service on her phone that charges a monthly membership fee of $10.00 and $0.5 per hour for the time spent listening to music. Which of the following functions gives Eva's cost, in dollars, for a month in which she spends h hours listening to premium music?

A) $f(x) = 10.5x$

B) $f(x) = 10x + 0.5$

C) $f(x) = 10 + 0.5x$

D) $f(x) = 10 + 50x$

What is the slope of the line in the xy-plane that passes through the points $\left(-\frac{7}{2}, -2\right)$ and $\left(-\frac{3}{2}, 1\right)$?

A) -1

B) $-\frac{2}{3}$

C) 1

D) $\frac{3}{2}$

20

A line is graphed in the xy-plane. If the line has a negative slope and a positive y-intercept, which of the following points cannot lie on the line?

A) $(5, 5)$

B) $(5, -5)$

C) $(-5, 5)$

D) $(-5, -5)$

21

Jason saves money each month to buy a new mobile phone. The total amount he has saved, T, can be calculated by the equation $T = 200 + 120m$, where m is the number of months since he started saving. What does the number 200 represent in the equation?

A) The amount of money Jason started with

B) The number of months Jason has been saving

C) The amount of money Jason saves each month

D) The total amount of money Jason wants to save

22

When the equation $y = 7x - 2c$, where c is a constant, is graphed in the xy-plane, the line passes through the point $(3, -3)$. What is the value of c?

A) 27

B) 3

C) –27

D) 12

23

The front of a rocket is now 150 feet above the ground. If the rocket rises at a constant rate of 26,000 feet per second after it is launched, which of the following equations gives height h, in feet, of the front of the rocket s seconds after it is launched?

A) $h = 26,000s + 150$

B) $h = 150s + 26,000$

C) $h = 433s + 150$

D) $h = 150s + 433$

24

If $f(x) = x - 3$ and $g(x) = 3x + 1$, what is the value of $g(4) - 4f(4)$?

An ant army parades at an average rate of 10 centimeters per second. Which of the following approximates the number of the distance, d, in centimeters, the ant army parades in t seconds?

A) $d = t + 10$

B) $d = t - 10$

C) $d = 10t$

D) $d = \dfrac{t}{10}$

Which of the following equations, when graphed in the xy-plane, would result in a line with slope of -3 that passes through the point $(0, 4)$?

A) $y = 3x + 4$

B) $y = 4x + 3$

C) $y = -3x + 4$

D) $y = -3x - 4$

Which of the following statements is true about the graph of the equation $3x - 5y = -3$ in the xy-plane?

A) It has a negative slope and a positive y-intercept.

B) It has a negative slope and a negative y-intercept.

C) It has a positive slope and a positive y-intercept.

D) It has a positive slope and a negative y-intercept.

LEVEL 2

`1`

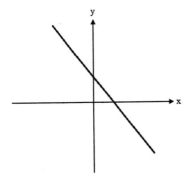

The graph of the function f is shown in the xy-plane above. The function f is defined by $f(x) = mx + b$, where m and b are constants. Which of the following could be the graph of the function g, where $g(x) = -mx - b$?

A)

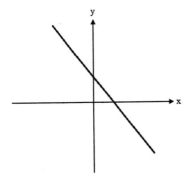

B)

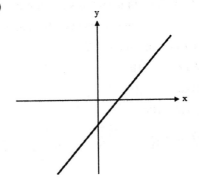

C)

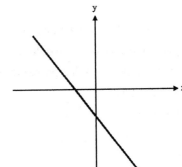

D)

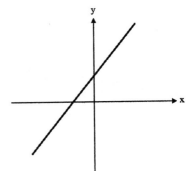

2

Nick opened a checking account with an initial deposit of x dollars, and then he deposited $35 into the account each week for 12 weeks. At the end of the 12 weeks, he had deposited $700 into the account. If no other deposits or withdrawals were made, what is the value of x?

A) 35

B) 65

C) 280

D) 700

3

In a remote town at China, 6 newcomers settle down for every 10 residents leaving this town. If 18 newcomers has settled down so far, how many people have left the town?

A) 20

B) 30

C) 40

D) 50

4

Arthur has recorded the height of a building being constructed, each month as it grew. The results are graphed below, and the line of best fit is also shown.

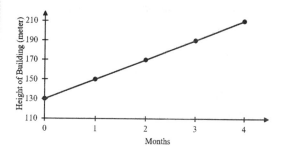

What is the meaning of the y-intercept of the graph?

A) The building is increasing by 130 m per month.

B) The building reached 130 m after the first month.

C) The building is increasing by 30 m per month.

D) The building is 130 m high when Arthur started measuring its height.

Hosea is an Uber driver. For each ride Hosea drive, he receives a commission of 15% of the amount by which the total fee is over $5. If the amount of order is d dollars, which of the following function gives him commission $C(d)$, in dollars, for each ride?

A) $C(d) = 0.15d$

B) $C(d) = 0.15(d - 5)$

C) $C(d) = 0.15(d + 5)$

D) $C(d) = 0.15(5 - d)$

The taxi fee in a remote town is 3$ per mile for the first 5 miles. After the first 5 miles, the payment increases to 5$ per mile. Which of the following functions gives the total taxi fee $F(m)$, in dollars, in terms of the number of miles driven, m, where $m > 5$?

A) $F(m) = 3(5) + 5m$

B) $F(m) = 3m + 5m$

C) $F(m) = 3m + 5(m - 5)$

D) $F(m) = 3(5) + 5(m - 5)$

The graph below gives the revenue of a bank from the time they opened till now. According to the graph, how many dollars of revenue they had at the very beginning when they opened the bank?

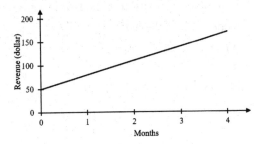

The value of a smart phone is $800. For every increase of two year, the value of the watch drops $100. Which of the following equations gives the approximate value d, in dollars, after y year?

A) $d = 800 - 100y$

B) $d = 800 - 50y$

C) $d = 800y$

D) $d = 100(400) - 2y$

Questions 9–10 refer to the following information.

Shanghai Health Department recorded the percentage of smoking people among the total population in Shanghai from 1980 to 2005. In the scatterplot below, *x* represents the number of years since 1980 and *y* represents the percentage of smoking people in Shanghai. The line of best fit for data is shown.

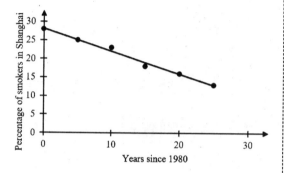

Which of the following is the best interpretation of the slope of line of best fit shown for the data?

A) The actual decrease in the percentage of smoking people each year

B) The predicted decrease in the percentage of smoking people each year

C) The actual decrease in the number of smoking people each year

D) The predicted decrease in the number of smoking people each year

9

Which of the following is closest to the equation of the line of best fit shown?

A) $y = 28x - 3$

B) $y = 28x + 3$

C) $y = -\frac{3}{5}x + 28$

D) $y = \frac{3}{5}x + 28$

An apartment already had 80 residents living inside. During the first year, 6 new residents are moving in the building each month, and 11 residents are moving out from the apartment each month. During the first year, which of the following equation best models the number of residents, r, at the apartment t months after.

A) $r = 80 + 6t$

B) $r = 80 - t$

C) $r = 80 + 5t$

D) $r = 80 - 5t$

A hotel is designing the arrangement of seating for a party. The figure below shows the first three possible arrangements of tables and maximum number of seats in each arrangement. If the number of table increases, and hotel is using the same design as in the graph, which of the following represents the maximum number of seats around n tables?

Arrangement 1

Arrangement 2

seat   table

Arrangement 3

A) $8n$

B) $2(4n + 2)$

C) $2(4n + 1)$

D) $2(4n + 4)$

Numbers	Total revenue
2	11
5	20
7	26
8	29

The values in the table above represents total revenue y, in dollars, to sell a kind of biscuits for number of bags x. The relationship between x and y is linear. If the data is graphed in the xy-plane, what is the slope of the line?

A) 5.5

B) 2

C) 3

D) 4

In the xy-plane, line m passes through two points $(0, 1)$ and $(2, 5)$, which of the following points lies on the line?

A) $(1, 3)$

B) $(3, 1)$

C) $(4, 10)$

D) $(8, 2)$

A block of ice is taken from the refrigerator and put into a heater and the switch was turned on. The initial temperature is -10°C, and the temperature rises 1°C every minute until it melts. Which of the following can be used to represent the temperature T, in Degree Celsius, of the ice after it is heated for t minutes?

A) $T(t) = t + 10$

B) $T(t) = t - 10$

C) $T(t) = -1 + 10t$

D) $T(t) = -1 - 10t$

The subscriber of a website y, in thousands after x days can be modeled as $y = 10,000 + 85x$. If the equation is graphed in the xy-plane, which of the following is the best interpretation of the y-intercept?

A) The number of subscribers increases by 85 every day.

B) There are 10,000 subscribers at the beginning.

C) The number of subscribers will be 10,000 in 85 days.

D) The number of subscribers will be 0 at the 85th day

As winter comes, the Danaus Plexippus (one kind of butterfly in a Mexican forest) fly to warmer place 20 km per day. The Mexican forest is 234 miles away from the warmer place. Which of the following function best approximates the distance $d(t)$, in km, of the distance they need to fly after t days, for $0 < t < 12$?

A) $d(t) = 234 - 20t$

B) $d(t) = 234 - 12t$

C) $d(t) = 234 + 20t$

D) $d(t) = 234 + 12t$

Questions 18–19 refer to the following information.

Between 2005 and 2018, data were collected every three years on the amount of plastic surgery proceeded annually in South Korea. The graph below shows the data and the line of best fit. The equation of the line of best fit is $y = 50x + 21$, where x is the number of years since 2005 and y is the amount of plastic surgery proceeded annually, in thousands.

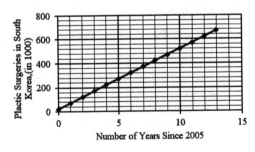

Which of the following is the best interpretation of the number 50 in the context of the problem?

A) The number of plastic surgeries, in thousands, proceeded in South Korea since 2005

B) The number of years South Korea took to do 1,000 plastic surgeries

C) The average annual decrease, in thousands, of plastic surgeries, proceeded per year in South Korea from 2005 to 2018

D) The average annual increase, in thousands, of plastic surgeries proceeded per year in South Korea from 2005 to 2018

Which of the following is closest to the percent increase in the number of plastic surgeries proceeded in South Korea from 2015 to 2018?

A) 29%

B) 35%

C) 42%

D) 53%

20

As Adele published her new album in 2015 in the first week, there were 23 million hits on the internet. Suppose the hits increases 200,000 every week, which of the following equations gives the approximate hits H, in million, after x weeks?

A) $H = 200,000x + 23$

B) $H = 2x + 23$

C) $H = 2x - 23$

D) $H = 0.2x + 23$

21

$$qx - 5y = 6$$
$$6x - 7y = 5$$

In the system of equations above, x and y are variables and q is a constant. For what value of q will the system of equations have no solution?

A) $-\dfrac{36}{5}$

B) $-\dfrac{70}{3}$

C) $\dfrac{30}{7}$

D) $\dfrac{36}{5}$

22

$$2x + 3y = 11$$
$$ax + by = 23$$

In the system of the equation above, a and b are constants. If the system has no solution, what is the value of $\dfrac{a}{b}$?

A) $\dfrac{2}{11}$

B) $\dfrac{23}{11}$

C) $\dfrac{2}{23}$

D) $\dfrac{2}{3}$

23

$$-5x - 7y = 1$$
$$2x + by = -0.4$$

In the system of equations above, b is a constant. For what value of b will be infinite many solutions to the system of equations?

A) $\dfrac{14}{5}$

B) $-\dfrac{14}{5}$

C) $\dfrac{8}{5}$

D) 2

$$20d + i = 260$$

An ice cream shop is running a promotion by giving a number of free ice cream samples each day. The equation above can be used to model the number of free ice cream samples, i, that remain to be given away d days after the promotion began. What does it mean that $(13, 0)$ is a solution to this equation?

A) During the promotion, 13 ice cream samples are given away each day.

B) It takes 13 days during the promotion to see 260 customers.

C) It takes 13 days during the promotion until none of the samples are remaining.

D) There are 13 samples available at the start of the promotion.

The equation $y = -10 + 0.05x$ models the relationship between the water level y, in meters, of a typical lake and the number of years, x, after it was measured in 2000. If the equation is graphed in the xy-plane, what is indicated by the y-intercept of the graph?

A) The age, in years, of the typical lake when it is measured in 2000

B) The water level, in meters, of the typical lake when it is measured in 2000

C) The number of years it takes the typical lake's water level to change

D) The number of meters the typical lake's water level changes each year

$$51x + 6 = 3(17x + c)$$

In the equation above, c is a constant. For what value of c does the equation have an infinite number of solutions?

Age	1	2	3	4	5
Height	80	84.8	90	95.1	100

The table above shows the height, in centimeters, of a boy. Which of the following best describes the type of model that fits the data in the table?

A) Linear, increasing by approximately 5 centimeters per year

B) Linear, increasing by approximately 10 centimeters per year

C) Exponential, increasing by approximately 5% each year

D) Exponential, increasing by approximately 10% each year

$$C = 75h + 130$$

The equation above gives the amount C, in dollars, a plumber charges for a job that takes h hours. Ms. Han and Mr. Joseph each hired this plumber. The plumber worked 2 hours shorter on Ms. Han's job than on Mr. Joseph's. How much less did the plumber charge Ms. Han than Mr. Joseph?

A) $75

B) $125

C) $150

D) $280

Line m in the xy-plane contains the point $(3, 5)$ and $(0, -1)$. Which of the following is an equation of line m?

A) $y = 2x + 1$

B) $y = 2x - 1$

C) $y = 3x - 1$

D) $y = 3x - 4$

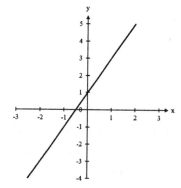

The graph of a line is shown in the xy-plane above, which of the following is an equation of the line?

A) $y = 2x + 1$

B) $y = 2x - 1$

C) $y = 3x - 1$

D) $y = 3x - 4$

31

$$F(d) = 1.5d + 10$$

The function F above models the swimming distance, in miles, of a student taking a swimming class, where d is the number of days the student has been in class and $0 \leq d \leq 15$. Accroding to the model, what is the swimming distance, in miles, of a student who had been in class for 6 days?

A) 11.5

B) 16.0

C) 19.0

D) 32.5

32

The graph of $y = f(x)$ is a line in the xy-plane that has a slope $\frac{4}{5}$. If $f(15) = 18$, which of the following functions could represent $f(x)$?

A) $f(x) = \frac{6}{5}x$

B) $f(x) = \frac{5}{4}x - 18$

C) $f(x) = \frac{4}{5}x + 18$

D) $f(x) = \frac{4}{5}x + 6$

33

$$T = 1.99c + 2.09n$$

The equation above gives the total cost T, in dollars, to purchase c pounds of cherries and n pounds of nectarines at a local grocery store. How much greater would be the total cost, in dollars, to purchase 5 pounds of cherries and 4 pounds of nectarines than to purchase 4 pounds of cherries and 3 pounds of nectarines? (Disregard the $ sign when gridding your answer. For example, if your answer is $4.97, grid 4.97)

$$x \le y + 2$$

$$3x - 2y \ge 6$$

In which of the following does the shaded region represent the solution set in the *xy*-plane to the system of inequalities above?

A)

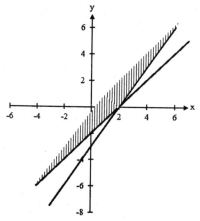

B)

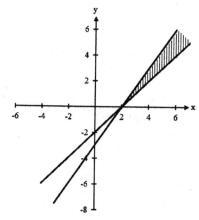

C)

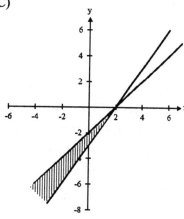

D)

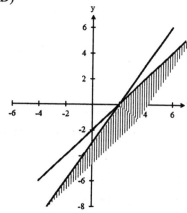

LEVEL 3

1

In the xy-plane, line l passes through the points $(-1,1)$ and $(-2, -5)$. Line m is perpendicular to line l and has a y-intercept that is 3 units less than the y-intercept of line l. What is an equation of line m?

A) $y = 6x + 7$

B) $y = 6x + 4$

C) $y = -\frac{1}{6}x + 7$

D) $y = -\frac{1}{6}x + 4$

2

For a service visit, Jack charges a \$39 fee plus an additional \$15 for every $\frac{1}{3}$ hour of work. If Jack's total charge was \$174, for how much time, in hours, did Jack charge?

Questions 3–4 refer to the following information.

In 2012, quarterly sales for full-service barber shops and hat stores in the United States each increased by a constant rate. For the first quarter of 2012, full-service barber shops had total sales of \$66,027 (in millions), and each quarter their sales increased approximately \$1,397 (in millions). For the first quarter of 2012, hat stores has total sales of \$5,143 (in millions), and each quarter their sales increased approximately \$22 (in millions).

3

Of the following equations, which best models the linear relationship in 2012 between the total quarterly sales (in millions), y, for full-service barber shops and the number of quarters, x, since the first quarter of 2012?

A) $y = 1,397x + 66,027$

B) $y = 1,397x - 66,027$

C) $y = 66,027x + 1,397$

D) $y = 66,027x - 1,397$

4

If the same linear trend for hat store sales continues, how many quarters after the first quarter of 2012 will the total quarterly sales (in million) for hat stores in the United States exceed \$6500?

A) 61

B) 62

C) 244

D) 248

The satisfaction level, s, of each successive measure of a customer who starts to take some bubble tea can be modeled by the equation $s = 2n + 30$, where n is the number of measures from the beginning when the customer starts to drink and $0 \leq n \leq 5$. According to the model, what is the change in satisfying level between each successive measure of the process?

A) 2

B) 30

C) 5

D) 10

The graph of lines l and m in the xy-plane are perpendicular. The equation of line m is $\frac{5}{6}x + \frac{2}{3}y = -5$. What is the slope of line l?

Oil is being drained from an oil tank at a constant rate. After four hours, 740 gallons of oil are left in the tank; After seven hours, 545 gallons are left. Which of the following functions best models $v(t)$, the volume of oil in the tank, in gallons, t hours after draining of the tank began?

A) $v(t) = 740 - t$

B) $v(t) = 740 - 65t$

C) $v(t) = 1000 - 195t$

D) $v(t) = 1000 - 65t$

The table shows data about the number of employees in XDF school.

	2016	2017	2018
Total employees	1,670	1,890	2,110
Percent male	35%	40%	45%
Percent female	65%	60%	55%

The total number of employees grows at a constant rate, and the percentages of male and female continue to decrease and increase by 5% respectively. How many male employees will be in XDF school in 2019?

$$\begin{cases} y > x \\ y \le -x + 3 \end{cases}$$

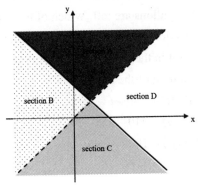

A system of inequalities and a graph are shown above. Which section or sections of the graph could represent all of the solutions to the system?

A) Section B

B) Section A and C

C) Section A and D

D) Section A, C and D

Tim climbs a hill that raise 5 meters in elevation for every 100 meters along the length of the hill path. The path is at 2,700 meters elevation where Tim began climbing the hill path, and he is climbing at 25 meters per minute along the path. What is the elevation of the path, in meters, at the point where Tim passes t minutes after beginning the climbing?

A) $2,700 + 0.05t$

B) $2,700 + 1.25t$

C) $2,700 + 5t$

D) $2,700 + 25t$

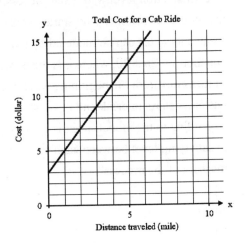

The line graphed in the xy-plane above models the total cost, in dollars, for a cab ride, y, in a certain city during peak hours based on the number of miles traveled, x. According to the graph, what is the cost for each additional mile traveled, in dollars, of a cab ride?

A) $2.00

B) $2.60

C) $3.00

D) $5.00

A company purchased a machine value at $30,000. The value of the machine depreciates by the same amount each year so that after 10 years the value will be $3,000. Which of the following equations gives the value, v, of the machine, in dollars, t, years after it was purchased for $0 \leq t \leq 10$?

A) $v = 30,000 - 2,700t$

B) $v = 3,000 + 2,700t$

C) $v = 30,000 - 3,000t$

D) $v = 3,000 + 30,000t$

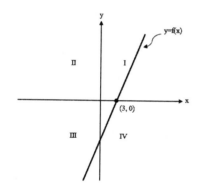

The graph in the xy-plane above shows the linear function f. If $g(x) = -2x + 4$, in which quadrant does the graph of $y = f(x)$ intersect the graph of $y = g(x)$?

A) I

B) II

C) III

D) IV

An agricultural festival charges a fixed cost per van for parking and an amount per person to enter the festival. One van with 3 people in it is charged $28, and another with 5 people is charged $40. How many people are in a van that is charged $82 to enter the festival?

$$3ax - 24 = 2(x - 9) + 7(x + 3)$$

In the equation above, a is a constant. If no value of x satisfies the equation, what is the value of a?

A) 1

B) 2

C) 3

D) 4

$$3x - 5y = 2$$
$$-1.5x + by = c$$

In the system of equations above, b and c are constants. If the system has infinitely many solutions, what is the value of c?

A) 1

B) –1

C) –2.5

D) 2.5

$$a(4x - 3) = 12x + 4$$

In the equation above, a is a constant. If no value of x satisfies the equation, what is the value of a?

A) 3

B) 2

C) 1

D) 0

5. Exponential Function

LEVEL 2

1

The population, in thousands, of a city is recorded every five years. The population of the city dropped at a constant rate as the time goes by. Which of the following graphs could represent the population of this city over time?

A)

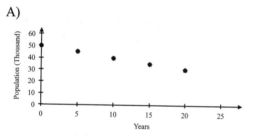

B)

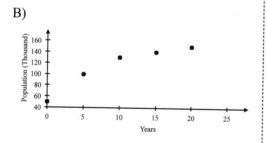

C)

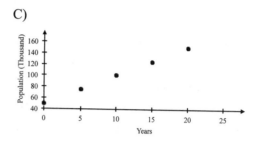

D)

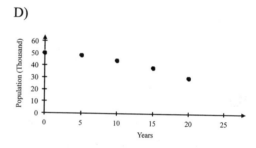

Which of the following describes an exponential relationship between the pair of variables listed?

A) The population of the town increases by 20% every 5 years

B) The salary of an employee who receives a $2000 increase in salary at the end of each year

C) The amount of rent for an apartment when the landlord raises the rent by $100 every 12 months

D) The amount of money in a retirement account that is decreasing in value by $4 each year

Which of the following does not describe an exponential relationship between the pair of variables listed?

A) Every 2 grams decreases m in the mass of the solution, the percent of the sugar contained increased by 1%.

B) Each second the number of the bacteria doubles.

C) Every day, the bees gather honey of 10 grams from flowers on average.

D) The velocity of the satellite triples each minute.

A particular element Po decays 16% every two years. If N_0 represents the remaining Po after t years, which of the following equation gives N_0?

A) $N_0 = N(0.16)^{\frac{t}{2}}$

B) $N = N_0(0.84)^{\frac{t}{2}}$

C) $N_0 = N(0.16)^t$

D) $N = N_0(0.84)^t$

The graph of the exponential function h in the xy-plane has a y-intercept of d, where d is a positive constant. which of the following could define the function h?

A) $h(x) = -3(d)^x$

B) $h(x) = 3(x)^d$

C) $h(x) = d(-x)^3$

D) $h(x) = d(3)^x$

1

For the years 1990 through 2006, the average number of batteries per person, b, not cycled in a year can be represented by the equation $b = 3.1(1.03)^t$, where t is the number of years after 1990. Which of the following is the best interpretation of the number 3.1 in the context of the problem?

A) The maximum number of batteries per person not recycled each year.

B) The increase in the number of batteries per person not recycled each year.

C) The average number of batteries per person not recycled in 1990.

D) The percent increase in the number of batteries per person not recycled each year.

2

A school district currently has 5,000 students and expects growth for the next 5 years. The school board in comparing two different growth projections. One model predicts that each year there will be 30 more students than the previous year. The other model predicts that each year there will be 1% more students than the previous year. To the nearest whole number, what is the difference in the projection of the total number of students these models predict after 5 years?

A) 20

B) 150

C) 105

D) 255

6. Quadratic Function

LEVEL 1

1

The function, $f(x) = x^2 + 4x - 21$ can be rewritten in the form $f(x) = (x + a)(x + b)$, where a is positive and b is negative. What is the value of a?

2

$$y = x^2 + 5x - 6$$
$$y = 3x + 2$$

The system of equations above is graphed in the xy-plane. Which of the following is the x coordinate of an intersection point (x, y) of the graphs of the two equations?

A) -4

B) 4

C) -3

D) -2

3

In the xy-plane, if the parabola with equation $y = ax^2 + bx + c$, where a, b, and c are constants, passes through the point $(-2, 2)$, which of the following must be true?

A) $2a - b = 1$

B) $-2b + c = 2$

C) $4a + 2b + c = 2$

D) $4a - 2b + c = 2$

4

Which of the following is a solution to the equation $x^2 - 3x = 1 + 3x^2 - 6x$?

A) -1

B) 0

C) 1

D) 2

5

Which of the following is equivalent to the expression $-3x^2 + 9x$?

A) $-3(x^2 + 3x)$

B) $-3x(x - 3)$

C) $x(3x + 9)$

D) $-x(x - 3)$

LEVEL 2

1

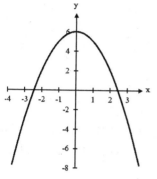

Which of the following equations best describes the figure above?

A) $y = -x^4 + 6$

B) $y = -(x^2 + 6)$

C) $y = -x^2 + 6$

D) $y = x^4 + 6$

2

$$4x(x - 1) + 5x = 9$$

Which of the following lists all solutions to the quadratic equation above?

A) $\frac{7}{8}, -\frac{3}{4}$

B) $\frac{-1 - \sqrt{145}}{8}, \frac{-1 + \sqrt{145}}{4}$

C) $\frac{-1 - \sqrt{145}}{8}, \frac{-1 + \sqrt{145}}{8}$

D) $\frac{-1 + \sqrt{37}}{8}, \frac{1 + \sqrt{37}}{8}$

3

The graph in the xy-plane of the following quadratic equations each have x-intercepts of 3 and –5. The graph of which equation has its vertex nearest to the x-axis?

A) $y = 2(x - 3)(x + 5)$

B) $y = -\frac{5}{7}(x - 3)(x + 5)$

C) $y = \frac{1}{3}(x - 3)(x + 5)$

D) $y = -4(x - 3)(x + 5)$

4

$$2x^2 - 5x - 3 = 0$$

If s and t are two solutions of the equation above and $s > t$, which of the following is the value of $s - t$?

A) $\frac{9}{2}$

B) $\frac{7}{2}$

C) $\frac{5}{2}$

D) $\frac{3}{2}$

5

Which of the following ordered pairs (x, y) satisfies both equations $y = 3x^2 - 2x + 1$ and $x = y - 7$?

A) (2, 9)

B) (1, 8)

C) (0, 7)

D) (0, 1)

6

If the equation $y = (x + 2)(x - 8)$ is graphed in the xy-plane, what is the y-coordinate of the parabola's vertex?

A) –25

B) –16

C) 16

D) 25

7

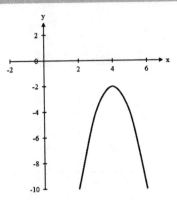

The graph of the function f in the xy-plane above is a parabola. Which of the following defines f?

A) $f(x) = -2(x - 4)^2 + 2$

B) $f(x) = 2(x - 4)^2 + 2$

C) $f(x) = 2(x - 4)^2 - 2$

D) $f(x) = -2(x - 4)^2 - 2$

8

If $f(x) = -3x^2 + 5$ and $f(x + a) = -3x^2 - 18x - 22$, what is the value of a?

A) –6

B) –3

C) 3

D) 6

9

$$h(x) = -12x^2 + 90x + 20$$

The quadratic function above models the height above the ground h in feet, of an arrow x seconds after it had been launched vertically. If $y = h(x)$ is graphed in the xy-plane, which of the following represents the real-life meaning of the positive x-intercept of the graph?

A) The initial height of the arrow

B) The maximum height of the arrow

C) The time at which the arrow reaches its maximum height

D) The time at which the arrow hits the ground

10

$$x^2 - 3x - 5 = 0$$

Which of the following are the solutions to the equation above?

A) $x = \dfrac{3}{2} \pm \sqrt{\dfrac{29}{4}}$

B) $x = \dfrac{3}{2} \pm \sqrt{\dfrac{29}{2}}$

C) $x = \dfrac{3}{4} \pm \sqrt{\dfrac{11}{4}}$

D) $x = \dfrac{3}{4} \pm \sqrt{\dfrac{11}{2}}$

11

$$y = 3x^2 - 5x + 4$$
$$y = 7x - 8$$

In the xy-plane, the graphs of the two equations above intersect at (a, b). What is the value of b?

A) 2

B) 4

C) 6

D) 8

12

$$2x^2 - 36x + 130 = 0$$

If $x = k$ represents a solution to the quadratic equation above, what is one possible value of k?

LEVEL 3

1

The function f is defined by $f(x) = 3x^2 - 18x + 27$. The graph of $y = f(x - h)$ is shown in the xy-plane below. What is the value of h?

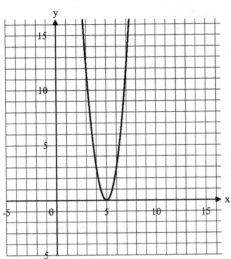

2

The graphs in the xy-plane of the following quadratic equations each have x-intercepts of 4 and 6. The graph of which equation has its vertex farthest from the x-axis?

A) $y = -20(x - 4)(x - 6)$

B) $y = 10(x - 4)(x - 6)$

C) $y = -\frac{1}{3}(x - 4)(x - 6)$

D) $y = -\frac{1}{2}(x - 4)(x - 6)$

3

In the xy-plane, if x_1, x_2, and $x_1 > x_2$ are two solutions to the equation $x^2 + 12x = 64$, what is the value of $x_1^2 - x_2^2$?

A) -120

B) -240

C) -360

D) -480

4

The length of a rectangle is 2 meters less than 3 times its width. What is the perimeter of the rectangular if the area of the rectangular is 2,133 square meters?

A) 212 meters

B) 210 meters

C) 200 meters

D) 186 meters

5

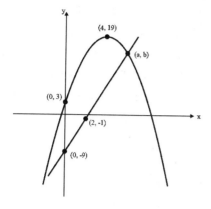

The *xy*-plane above shows one of the two points of intersection of the graphs of a linear function and a quadratic function. The shown point of intersection has coordinates (a, b). If the vertex of the graph of the quadratic function is at $(8, 19)$, what is the value of a?

Questions 6–7 refer to the following information.

$$v = v_0 - gt \text{ (speed-time)}$$

$$h = v_0 t - \frac{1}{2} g t^2 \text{(position-time)}$$

$$v^2 = v_0^2 - 2gh \text{ (position-speed)}$$

An apple is thrown upward with an initial speed of 20 meters per second (m/s). The equations above describe the constant-acceleration motion of the apple, where v_0 is the initial speed of the apple, v is the speed of the apple as it is moving up in the air, h is the height of the apple above the ground, t is the time elapsed since the apple was projected upward, and g is the acceleration due to gravity (9.8 m/s^2).

6

How long will it take for the apple to reach its maximum height to nearest second?

7

What is the maximum height from the ground the apple will rise to the nearest meter?

8

$$\frac{x}{x-4} = \frac{3x}{3}$$

Which of the following represents all the possible value of x that satisfy the equation above?

A) 0 and 3
B) 0 and 5
C) 3 and 5
D) 5

If $y = 4x^2 - 8x + 3$ is graphed in the xy-plane, which of the following characteristics of the graph is displayed as a constant or coefficient in the equation?

A) y-coordinate of the vertex

B) x-intercept(s)

C) y-intercept

D) x-intercept of the line of symmetry

The function f is defined by $f(x) = -ax^2 + bx + 5$, where a and b are positive constants. Which of the following points CANNOT be on the graph of $y = f(x)$ in the xy-plane?

A) $(-1, 6)$

B) $(0, 5)$

C) $(4, 0)$

D) $(5, -3)$

The height h, in feet, of a toy rocket above the ground t seconds after it is launched can be modeled by the function $h(t) = -16t^2 + 224t$. Which of the following methods can be used to find m, the maximum height, in feet, of the rocket above ground?

A) Compute $h(0) = -16 \times (0)^2 + 224 \times (0)$ to find $h(0) = m$.

B) Rewrite h as $h(t) = -t(16t - 224)$. The number 224 represents m.

C) Rewrite h as $h(t) = -16t(t - 14)$. The number 14 represents m.

D) Rewrite h as $h(t) = -16(t - 7)^2 + 784$. The number 784 represents m.

7. Function

LEVEL 1

1

$$y = -2x + 7$$

Given the equation above, if $x = 3$, what is the value of y?

2

Which of the following equals
$6x^2 + 2x(x - 3) + 7(2x + 1) - 5$?

A) $8x^2 + 2x + 5$

B) $8x^2 + x + 2$

C) $4x^2 + 6x + 2$

D) $8x^2 + 8x + 2$

3

$$f(m) = \frac{4m + 2x}{5} + 100$$

The number of bees coming in hundreds every summer visiting the flowers in Yellowstone Park can be modeled by the function above, where x is a constant and m is the month of a year in 1-12. If 4,100 bees visit flowers in the Yellowstone park, when it is in July, which is the value of x?

A) 9,986

B) 7,345

C) 1,088

D) 8,744

4

If $f(x) = x^2 + 2x + 7$ and x is a positive integer number less than 2, what is the value of $f(x)$?

LEVEL 2

1

$$C_{(x)} = \frac{2(5t - 3a)}{3} + 10$$

The number of customers who go to a ski resort can be modeled by the function $C_{(x)}$, where a is a constant and x is the air temperature in Celsius degree (°C) for $0C° < t < 35C°$. If 40 customers are predicted to go to the ski resort when the temperature is 12C°, What is the value of a?

A) 5

B) 10

C) 15

D) 20

2

If $f(x) = x^2 - 2x + 5$ and b is a positive integer less than 4, what is one possible value of $f(b)$?

3

The function f is defined as $f(h) = (h - 2)^2 (h - 3)$. If $f(a + 2) = 0$, what is one possible value of a?

4

A purchasing manager of a restaurant purchased 100 kg of food from the market for a total of $8,000. Some of the food was beef which worth $200 per kilogram, and the rest was pork which worth $50 per kilogram. How many more kilograms of pork did the manager purchase than beef?

5

If $f(x) = x + 5$, for what value of a is $f(3a) + 2 = f(2a) + 3$?

6

If $h(x) = 2^{x^2+3x-5}$, what is the value of $h(2)$?

A) 16

B) 8

C) 32

D) 64

Questions 7–8 refer to the following information.

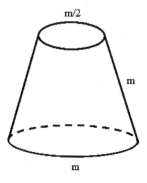

The glass pictured above can hold a maximum volume of 576 cubic centimeters, which is approximately 21 fluid ounces.

$$Volume = \frac{7\pi}{48}m^3$$

7

What is the value of m, in centimeters?

A) 2.79

B) 6.54

C) 2.54

D) 10.79

8

Water pours into the glass at a constant rate. Which of the following graphs best illustrates the height of the water level in the glass as it fills?

A)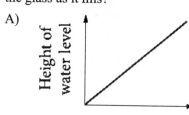

B)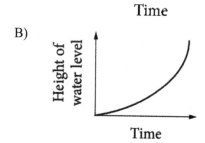

C)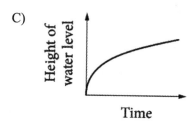

D)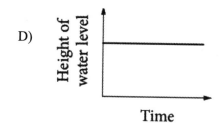

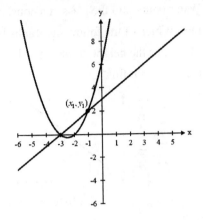

The two equations are graphed in the *xy*-plane above. Which of the following could be a solution (x_1, y_1) to this system of equation?

A) $(-1, 1)$

B) $(-1, 2)$

C) $(3, 0)$

D) $(-3, 1)$

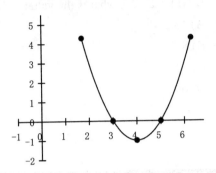

What is the maximum value of the function graphed on the *xy*-plane above, for $2 \leq x \leq 6$?

A) 3

B) 2

C) 0

D) –1

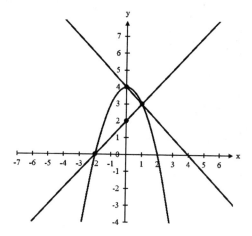

A system of three equations is graphed in the *xy*-plane above. How many solutions does the system have?

A) None

B) One

C) Two

D) Four

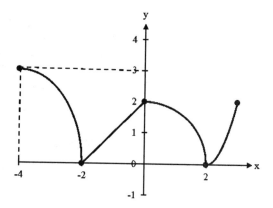

The figure above shows the complete graph of the function *f* in the *xy*-plane. The function *g* (not shown) is defined by $g(x) = f(x) + 6$. What is the maximum value of the function *g*?

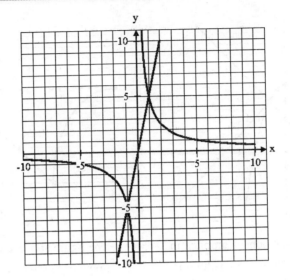

The graph of two equations a system of equations are shown in the *xy*-plane above, which of the following statements is true?

A) The point (0, 0) represents the solution to the system.

B) The point (–1, –5) and (1, 5) represent the solution to the system.

C) The point (5, 5) represents the solution to the system.

D) The point (5, 0) and (0, 5) represents the solution to the system.

LEVEL 3

1

The function f is defined by $f(x) = (x-3)^2$.
If $f(x+k) = x^2 + 8x + 16$, where k is a
constant, what is the value of k?

A) 7

B) 4

C) 1

D) –4

2

In the xy-plane, the graph of $y = 2(x-3) - 2$
is the image of the graph of $y = 2(x+5) - 2$
after a translation of how many units to the
right?

3

In the xy-plane, the graph of $y = 2x + 3$ is
the image of graph $y = 2x$ after a translation
of how many units to the left?

4

If $H(5 - 2x) = \sqrt{x^2 + 3x + 5}$ for all real
numbers x, what is the value of $H(3)$?

5

If $f(x + 2) = 3x - 1$ for all values of x, what
is the value of $f(1)$?

A) –4

B) 2

C) 5

D) 8

8. Polynomial Factors and Graphs

LEVEL 2

1

The polynomial $p^3 - p^2 - 4p + 4$ can be written as $(p^2 - 4)(p - 1)$. What are all of the roots of the polynomial?

A) 2, 1

B) –2, 1

C) 2, –2, 1

D) 1, –1, 2

2

$$f(x) = (x + 5)(x^2 + 6x + 5)$$

How many distinct zeros does the function f, defined?

A) None

B) One

C) Two

D) Three

3

Which of the following could be the graph of $y = (x + 1)(x - 2)(x + 3)$?

A)

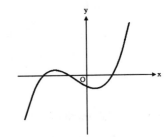

B)

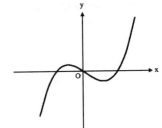

C)

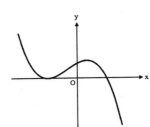

D)

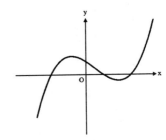

LEVEL 3

1

$$\frac{1}{3x-1} + 4$$

Which of the following is equivalent to the expression above for $x > 0$?

A) $\dfrac{4x+1}{3x-1}$

B) $\dfrac{4x-3}{3x-1}$

C) $\dfrac{12x-3}{3x-1}$

D) $\dfrac{12x+1}{3x-1}$

2

In the xy-plane, the graph of the polynomial function f crosses the x-axis at exactly two points, $(a, 0)$ and $(b, 0)$, where a and b are both positive. Which of the following could define f?

A) $f(x) = (x-a)^2(x-b)$

B) $f(x) = (x+a)(x+b)^2$

C) $f(x) = (x-a)^3(x+b)^2$

D) $f(x) = x(x-a)(x+b)$

9. Unit

LEVEL 1

1

Cheetah is the fastest animal in the world and it can run at a speed of 71 miles per hour. How fast can a cheetah run in kilometers per hour, rounded to the nearest 100 kilometers per hour? (1 mile is approximately 1.61 kilometers)

A) 114

B) 120

C) 44

D) 45

2

A pen costs p dollars, and a notebook cost 1 dollar more than a pen. If 2 pen and 3 notebooks cost 83 dollars, what is the value of p?

A) 15

B) 16

C) 17

D) 18

3

Two units of length used in ancient Egypt were cubits and palms, where 1 cubit is equivalent to 7 palms. The Great Pyramid of Giza is approximately 280 cubits long. Which of the following best approximates the length, in palms, of the pyramid?

A) 0.1

B) 40

C) 280

D) 1960

4

Biologists found a new species of dragonfish at the world's deepest undersea vent, the Beebe Vent Field. The vent is 3.1 miles below the sea's surface. Approximately how many kilometers below the sea's surface is the vent? (1 kilometer is approximately 0.6214 miles)

A) 2

B) 3

C) 4

D) 5

At a certain point in time, 1 US dollar was equivalent to 6.87 RMB yuan. Based on this relationship, what was the value of 100 RMB yuan in dollars at that time?

A) 687.00

B) 14.56

C) 106.87

D) 93.13

LEVEL 2

The average house price in Shanghai is $¥9 \times 10^4$ per square meter. Eric just bought a house which worth $¥2.25 \times 10^7$ in total, what is the approximate total area, in square meter, of the house?

A) 50

B) 100

C) 150

D) 250

Peter is driving a car at a constant speed of 100 kilometers (km) per hour. Which of the following is closest to the total distance of Peter's drive, in miles, over the course of 3 hour time period? (1km=0.62mile)

A) 62

B) 186

C) 300

D) 100

A bubble tea house purchases 50 bags of bubbles with 20 bubbles per bag. Each cup of bubble tea is made by mixing 10 bubbles with milk. What is the maximum number of bubble tea that can be made from 50 bags of bubbles?

A) 50

B) 80

C) 100

D) 200

The speed of sound in air at 20°C is about 767 miles per hour. What is the speed, rounded to the nearest whole number, in meters per second? (1mile=1,609 meters)

LEVEL 3

1

A doctor prescribes a medication for a 7716-pound elephant. In each dose, the elephant is given 0.1 milligrams of the medication for each kilogram of the elephant's body weight. The medication is given to the elephant twice per day for 2 days. How much medication, to the nearest milligrams, should be provided for the entire 2-day treatment plan? (1 kilogram = 2.2 pounds)

2

The wheel of a bike spins 50 revolutions per minute and its radius is 50 centimeters. At this rate, how far will it travel in 2 hours (in km)?

A) 6π

B) 60π

C) 600π

D) 1200π

10. Percent

LEVEL 1

1

The price of latest iPhone is $900. The price of the next model would increase by 10%. How much would the next model of iPhone be, in dollars?

A) 905
B) 910
C) 990
D) 1350

2

The scientists have found that the number of bats in a cave has decreased 10%. Last time, the recorded the number was 87,230, which of the following is the number of bats in that cave this time?

A) 78,507
B) 78,508
C) 95,953
D) 95,931

3

Echo has a transport card with an initial value of ¥233. Each time she takes the subway, ¥3.6 is subtracted from the value of the card. What percent of the initial value of Echo's card are the two subway rides?

A) 1.5%
B) 3.1%
C) 2.7%
D) 1.4%

LEVEL 2

1

A university has 200 Chinese, 50 Indians and 250 Mexicans students. What percent of this university are Indians?

A) 5%

B) 10%

C) 15%

D) 50%

2

Robert purchased a laptop that cost $3,000 before a 5% sales tax. How much did Robert pay in total, for the laptop? (Disregard the $ sign when gridding your answer. For example, if your answer is $500, grid 500)

3

Susan has a membership card for a gym that had an initial value of $95. For every time Susan uses gym, $2, the cost of gym, is subtracted from the value of the membership card. What percentage of the initial value of Susan's membership card is the cost of gym for each use?

A) 2%

B) 2.1%

C) 2.2%

D) 2.3%

4

Echo earned $60,000 in 2016. In 2017, the salary she earned decreased by 10% percent, and in 2019, the money she earned increased by 20% percent more than that in 2017. How much does she earn in 2018?

A) 79,200

B) 792,000

C) 64,800

D) 52,800

5

A university enrolled 100 females and 200 males last year. This year, the university enrolled 40 percent more females and 30 percent fewer males compared to last year. By approximately what percentage did the number of students in the university decline?

A) 70 percent

B) 10 percent

C) 7 percent

D) 5 percent

LEVEL 3

The population of Town A grew by approximately 27% between 1996 and 2000 and then decreased by approximately 25% between 2000 and 2004. If the population was 36,400 in 2000, what was the approximate change in the number of people living in Town A between 1996 and 2004?

A) Increase of 1361 people

B) Decrease of 1361 people

C) Increase of 2555 people

D) Decrease of 2555 people

11. Proportion/Fraction/Probability

LEVEL 1

1

In a remote town at China, 6 newcomers settle down for every 10 residents leaving this town. If 18 newcomers has settled down so far, how many people have left the town?

A) 20

B) 30

C) 40

D) 50

2

In 2018, there are about 19,800 teachers working in New Oriental. About 20% percent of the teachers teach TOEFL, and the ratio of male to female teachers is about 1:2, how many female teachers teaching TOEFL in New Oriental in 2018?

3

	Music style			
	Classic	Pop	jazz	total
Class A	5	72	34	111
Class B	43	11	23	77
Total	48	83	57	188

The table above shows the students' preference for the music styles. What fraction of the students in Class A enjoy pop music, to the nearest hundredth?

LEVEL 2

1

	Favorite sport				
	Baseball	Basketball	Football	Others	Total
Freshmen	30	50	20	50	150
Sophomore	50	10	50	40	150
Junior	100	10	20	70	200
Senior	30	20	20	80	150
Total	210	90	110	240	650

The table above shows the number of students' favorite sports. Based on the table, what proportion of students in Senior year chose baseball as their favorite sport?

A) 0.15   B) 0.18   C) 0.2   D) 0.25

2

Distribution of 50 Pairs of Sneakers

Brand	Basketball shoes	Baseball shoes
Nike	15	20
Adidas	10	10

Morgan has a collection of 55 pairs of sneakers from Nike and adidas. The distribution of the sneakers is shown in the table above. If a pair of basketball shoes is to be selected at random, what is the probability that the brand is Nike?

A) $\frac{3}{5}$ B) $\frac{2}{5}$ C) $\frac{1}{3}$ D) $\frac{2}{3}$

Cupcakes are given to Oleg, Earl and Han in the ratio of 2:5:3. If Han gets 12 cupcakes, how many cupcakes did Earl get?

A) 14

B) 18

C) 20

D) 16

John collected a number of species A and B by sampling. His records are shown in the table below.

		Species B	
		Present	Absent
Species A	Present	6	15
	Absent	20	9

For what percent of all the data did species A present?

A) 12%

B) 42%

C) 29%

D) 37%

Bottles of Juice of a Supermarket

	Brand A	Brand B
Apple juice	212	340
Orange juice	432	157

The table above presents information about the storage of apple juice and orange juice of a supermarket. Approximately what percentage of the bottles of juice from Brand A are orange juice?

A) 41 percent

B) 49 percent

C) 67 percent

D) 73 percent

6

New Oriental High School randomly selected freshman, sophomore, junior, and senior students for a survey about potential changes to next year's schedule. Of students selected for the survey, $\frac{2}{5}$ were freshmen and $\frac{1}{3}$ were sophomores. Half of the remaining selected students were juniors. If 480 students were selected for the survey, how many were seniors?

A) 64

B) 160

C) 192

D) 240

Questions 7–8 refer to the following information.

The table below compares features of the orbits of Earth, Pluto and Jupiter

	Earth	Pluto	Jupiter
Semimajor axis (in 10^6 kilometers)	149.60	5,906.38	778.57
Perihelion (in 10^6 kilometers)	147.09	4,436.82	740.52
Aphelion (in 10^6 kilometers)	152.10	7,365.93	816.62

7

The ratio of Pluto's semimajor axis to Earth's semimajor axis is $p:1$, where p is a constant. What is the value of p, rounded to the nearest tenth?

8

To the nearest tenth, what percent of Jupitor's perihelion distance is Earth's perihelion distance? (Disregard the % sign when gridding your answer)

LEVEL 3

1

Phillip performed an experiment where he recorded the weather in two different cities, Shanghai and Beijing, to see whether it rained or not. His results are shown in the table below.

		Shanghai	
		Rainy	Not rainy
Beijing	Rainy	36	18
	Not rainy	24	22

What percent of the days in Shanghai are rainy?

A) 40%

B) 46%

C) 54%

D) 60%

2

Nation and Age of Football Players of Barcelona

	Age			
	16-22	23-28	29-36	Total
Europe	13	15	8	36
Africa	7	9	3	19
America	10	11	9	30
Asia	2	3	1	6
Total	32	38	21	91

The table above shows the distribution of football players according to their nations and age range. Based on the information in the table, what fraction of players who are not from Europe aged 23-28?

A) $\frac{23}{55}$

B) $\frac{38}{55}$

C) $\frac{15}{36}$

D) $\frac{9}{36}$

A survey of 200 randomly selected senior high school students aged 15 to 18 in the United States was conducted to gather data on preference between basketball and football. The data are shown in the table below.

	Preferring basketball	Preferring football	Total
Ages 15-16	31	44	75
Ages 17-18	73	52	125
Total	104	96	200

3

Which of the following is closest to the percent of those surveyed who prefer basketball?

A) 15%

B) 36%

C) 52%

D) 62%

4

In 2016 the total population of students in the United States who were between 15 and 18 years old (inclusive) was about 20 million. If the survey results are used to estimate information about preferring basketball or football across the country, which of the following is the best estimate of the total number of students between 15 and 16 years old in the United States who prefer football in 2016?

A) 9,600,000

B) 4,400,000

C) 3,100,000

D) 440,000

Questions 5–7 refer to the following information.
A survey is conducted to see the preference of tea or coffee among 69 employees in a company. Each employee makes a choice between coffee and tea. A summary of the survey is shown in the table below.

Preference of Coffee or Tea for
the 69 Employees

Age	Coffee	Tea
30 or greater	26	17
Less than 30	12	14

What is the difference, to the nearest whole percent, between the percentage of the employees prefer coffee whose age is 30 or greater and the percentage of employees prefer tea whose age is 30 or greater?

A) 3%
B) 10%
C) 14%
D) 23%

If an employee of the company is chosen at random, which of the following is closest to the probability that the employee's age will be 30 or greater?

A) 0.62
B) 0.38
C) 0.45
D) 0.55

Of the employees who prefer coffee, the ratio of women to men is approximately 1:4. Which of the following is the best estimate of the number of men who prefer coffee in the survey?

A) 6
B) 8
C) 25
D) 30

Customer Purchase at a Gas Station

	Beverage purchased	Beverage not purchased	Total
Gasoline purchased	52	32	84
Gasoline not purchased	33	27	60
Total	85	59	144

On Tuesday, a local gas station had 144 customers. The table above summarizes whether or not the customers on Tuesday purchased gasoline, a beverage, both, or neither. Based on the data in the table, what is the probability that a gas station customer selected at random on that day did NOT purchase gasoline?

A) $\frac{27}{60}$

B) $\frac{27}{59}$

C) $\frac{33}{60}$

D) $\frac{60}{144}$

	Female	Male	Total
Democrat	78	190	268
Republican	25	265	290
Independent	2	2	4
Total	105	457	562

The table above shows a summary of the members of the 123th Congress of the United States in November 2014, categorized by gender and political affiliation. If one of the Democrats in the 123th Congress is selected at random, which of the following is closest to the probability that the member of Congress is female?

A) 18%

B) 29%

C) 74%

D) 80%

12. Scatterplot

LEVEL 2

1

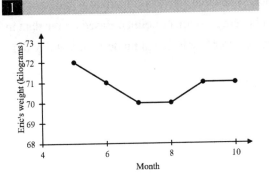

Month

The graph above is the fluctuation of Eric's weight, in kilograms, from May to October. From May to July, how many kilograms did Eric lost each month?

A) 2

B) 1

C) $\frac{1}{2}$

D) $\frac{1}{4}$

Questions 2–3 refer to the following information.

Rick is investigating the relationship between the price of houses per square meter in Shanghai and the number of subway stations nearby. He selected 22 houses in Shanghai at random. For each house, he records the number of nearby subway stations and the price of each house per square meter. The results are shown, along with the line of best fit, in the scatterplot below.

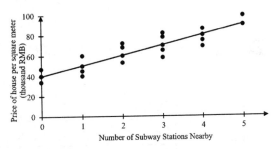

Number of Subway Stations Nearby

2

A house in the research worth 60,000 RMB per square meter. If this is less than that predicted by the line of best fit, what is the least number of nearby subway station this house could have?

A) 1

B) 2

C) 3

D) 4

The line of best fit passes through the point (–2, 20000). Which of the following can be concluded from this?

A) The line of best fit will not model the number of subway stations nearby with low price per square meter.

B) A house with 20000 RMB per square meter can not have subway station nearby.

C) A house with 20000 RMB per square meter do not exist.

D) The house price per square meter in Shanghai can not be lower than 20000 RMB.

Questions 4–7 refer to the following information.

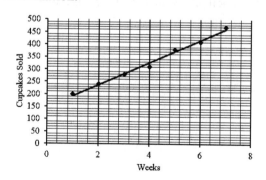

The above graph shows the sales of Max's homemade cupcakes for the past 7 weeks. According to the best fit line, of the following, which best predicts the sales of cupcakes in the 8th week?

A) 464

B) 763

C) 892

D) 514

What percent of the weeks did Max sell more than 400 cupcakes?

A) 28.6%

B) 31.2%

C) 35.7%

D) 42.1%

Of the following, which of the data can best estimate the median sales of the first 4 weeks?

A) 242

B) 250

C) 260

D) 285

7

Max predicted that the sales in the 10th week would be 780, which of the following best describes Max's prediction?

A) It is about 100 greater than the best fit line predicted.

B) It is about 100 less than the best fit line predicted.

C) It is about 150 greater than the best fit line predicted.

D) It is about 150 less than the best fit line predicted.

8

In the 2000s, the park rangers at Yellowstone National Park implemented a program aiming at increasing the dwindling wolf population in Montana. Results of studies of the wolf population in the park are shown in the scatterplot below.

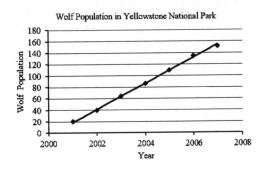

Wolf Population in Yellowstone National Park

Based on the line of best fit in the scatterplot above, which of the following is the closest to the average annual increase in wolf population in Yellowstone National Park between 2001 and 2006?

A) 22   B) 28   C) 34   D) 40

Questions 9–10 refer to the following information.

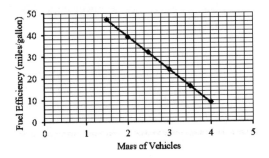

9

The scatterplot above shows the fuel efficiency, in miles per gallon, of a variety of vehicles weighing between 1.5 and 4 tons. Based on the line of best fit to the data, which of the following is the closest to the expected fuel efficiency, in miles per gallon, of a vehicle that weighs 3 tons?

A) 20

B) 24

C) 27

D) 28

The line passes through (6, –22), which of the following can be concluded from this?

A) The line of the best-fit line will not model the efficiency well for a vehicle with a heavy mass.

B) The vehicle with 6 tons of mass can no longer take fuels any more.

C) The vehicle with 6 tons of mass cannot decrease its efficiency any further.

D) A vehicle can not have mass of 6 tons.

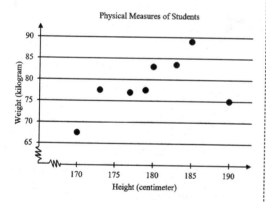

Physical Measures of Students

The scatterplot above shows the weights and the heights of 8 students. What is the weight, in kilograms, of the student represented by the data point that is farthest from the line of best fit (not shown)?

A) 89

B) 83

C) 75

D) 67

Which scatterplot shows a negative association that is linear? (Note: A negative association between two variables is one in which higher values of one variable correspond to lower values of the other variable, and vice versa.)

A)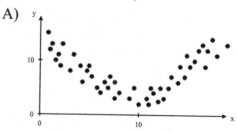

B)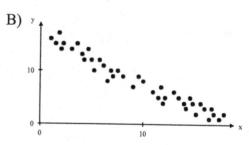

C)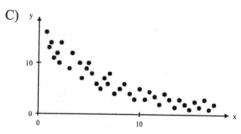

D)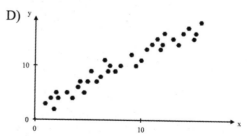

A university counselor conducted a study over 24 consecutive months to determine the number of students with boyfriends or girlfriends. Each student in the 2016 graduating class is surveyed once per month for the first two years of university. The graph below shows the data for each month the survey was conducted.

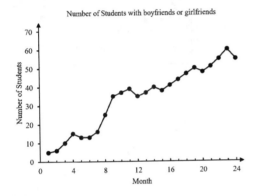

Number of Students with boyfriends or girlfriends

During which of the following periods is the increase in the number of students with boyfriends or girlfriends largest?

A) Months 2 through 5
B) Months 7 through 10
C) Months 14 through 17
D) Months 20 through 23

Questions 14–15 refer to the following information.

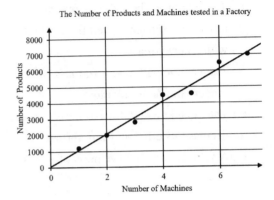

The Number of Products and Machines tested in a Factory

The scatterplot above shows the number of products and the number of machines tested in a certain factory in seven different days. The line of best fit for the data is also shown.

Which of the following statements about the relationship between the number of products and the number of machines is true?

A) As the number of times at bat increases, the number of hits decreases.
B) As the number of times at bat increases, the number of hits increases.
C) As the number of times at bat increases, the number of hits remains constant.
D) As the number of times at bat decreases, the number of hits increases.

For the day when there are 3 machines in the factory, the actual number of products that day is approximately how many fewer than the number of products predicted by the line of best fit?

A) 20

B) 200

C) 500

D) 1000

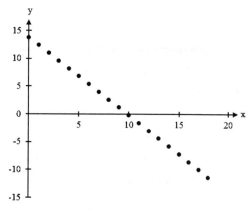

The population y, in tens of thousands, of a certain city x years later since 1990 is shown in the scatterplot below. If the data are modeled using a line of best fit, which of the following could be the equation of that line?

A) $y = 13.8 + 1.4x$

B) $y = 13.8 - 1.4x$

C) $y = 12.0 + 1.4x$

D) $y = 12.0 - 1.4x$

LEVEL 3

Questions 1–3 refer to the following information.

A history teacher did a survey on the relation between the time that each student spent on the course and his score in the final exam. The survey data of all 12 students are shown below.

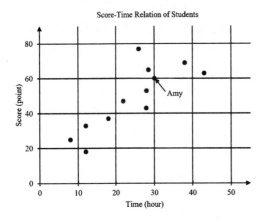

The data is modeled by the equation $S = 2T$, where S is the score, in points, each student got in the final exam of the course and T is the time, in hours, he spent on the course.

1

According to the data above, how much time, in minutes, did Amy spend on the course?

A) 30 minutes

B) 1800 minutes

C) 2400 minutes

D) 3600 minutes

2

There are three students shown in the graph spent approximately 28 hours on the course. Which of the following is closest to the range of scores of these three students, in points?

A) 35

B) 25

C) 15

D) 5

3

Based on the model, if a student spent 45 hours on the history course, what would be the score, in points, of this student?

A) 9 points

B) 45 points

C) 90 points

D) 900 points

84

Questions 4–5 refer to the following information.

In an experiment, a heated cup of coffee is removed from a heat source, and the cup of coffee is then left in a room that is kept at a constant temperature. The graph above shows the temperature, in degrees Fahrenheit (°F), of the coffee immediately after being removed from the heat source and at 10-minute intervals thereafter.

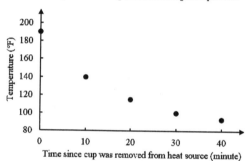

Temperature of a Cup of Coffee during an Experiment

4

Of the following, which best approximates the temperature, in degrees Fahrenheit, of the coffee when it is first removed from the heat source?

A) 75
B) 100
C) 135
D) 190

5

During which of the following 10-minute intervals does the temperature of the coffee decrease at the greatest average rate?

A) Between 0 and 10 minutes
B) Between 10 and 20 minutes
C) Between 20 and 30 minutes
D) Between 30 and 40 minutes

13. Box Plot

LEVEL 1

1

The box plot below shows the average score earned per game of the members of a school basketball team. What is the range of average score?

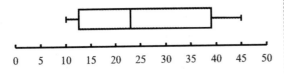

A) 13
B) 22
C) 26
D) 35

LEVEL 2

1

The following box plots show the distribution of weight, in kilograms (kg) of two species of pig.

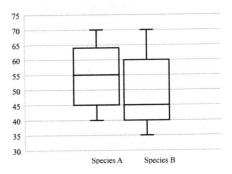

What is the approximate difference, in kg, between the range of weight of Pig A and the range of weight of Pig B?

A) 0
B) 5
C) 10
D) 15

LEVEL 3

Questions 1–2 refer to the following information.

Ushio wanted to determine if there is difference between the GPA of his male and female students, so he conducted a survey to collect the GPA as well as the gender of his students. Each student's GPA is rounded to the nearest 0.5. The results of boys and girls are summarized in the statistical graphs below.

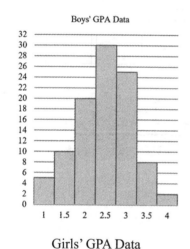

Boys' GPA Data

Girls' GPA Data

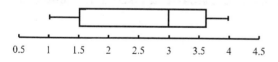

1

If B represents the median GPA of boys and G represents the median GPA of girls, what is the value of $G-B$?

A) 3

B) 2

C) 0.5

D) 0

Which of the following box plots can represent the GPA data of boys?

A)

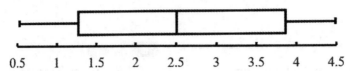

B)

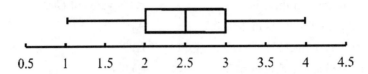

C)

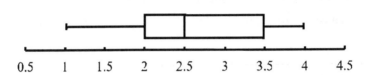

D)

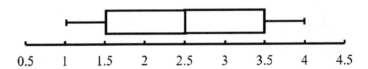

14. Statistics

LEVEL 1

Questions 1–2 refer to the following information.

Volume of different liquids for one cup of standard cocktail sold in a lounge bar

Cocktails	Volume (ml)	Total Alcohol (ml)	Water (ml)
Gin and tonic	300	100	100
Long Island	350	130	50
Mojito	250	80	50
Old Fashioned	250	150	100
Manhattan	300	140	120

The table above shows the total volume of cocktails, in milliliter, and volume of alcohol and water for each cocktail.

1

How much more liters of water are in one cup of Manhattan than in one cup of Long Island? (1 liter=1000 milliliter)

A) 5

B) 70

C) 0.05

D) 0.07

2

Gordon ordered 1 cup of each of the five cocktails in the bar. Of the following, which best approximates the mean number of volumes, in liters, of per cup of cocktail? (1 liter=1000 milliliter)

A) 290

B) 350

C) 0.35

D) 0.29

Monthly salary of 50 employees in Alpha company

Age group	Number of employees	Average monthly salary
Under 35	12	$3100
35-50	24	$3800
Over 50	14	$4200

The chart above shows the number of employees in Alpha company in three age groups and the average monthly salary of the workers in each group. What is the average monthly salary, in dollars for all the employees?

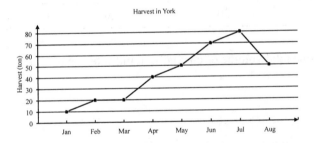

Harvest in York

The line graph above shows the monthly harvest of kinds of crops from January to August last year in York. According to the graph, what was the greatest change (in absolute value) in the monthly harvest between two consecutive months?

A) 10 tons

B) 20 tons

C) 30 tons

D) 40 tons

LEVEL 2

Means of transportation	Number of students				
Bike	20	5	10	0	0
Car	10	0	15	15	10

Table above shows the number of students who choose different means of transportation. The number of students of 5 surveys has been shown in the table. What is the positive difference between the average of those who choose bike and those who choose car as their means of transportation?

A) 15

B) 5

C) 4

D) 3

Heights of 16-year-olds (in inches)

68	68	70	69	71
73	57	72	69	69
71	77	70	68	66

The table above lists the heights, to the nearest inch, of a random sample of 15 16-year-olds at school. Which of the following values will change the least if 57-inch measurement is removed from the data?

A) They will all change by the same amount

B) Range

C) Mean

D) Median

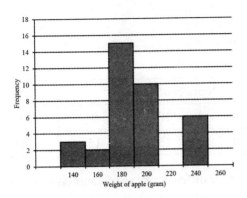

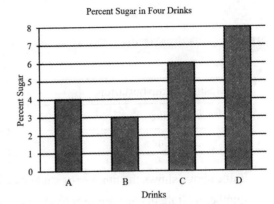

The histogram above shows the distribution of the weights, in grams, of 36 apples in a basket. Which of the following could be the median weight of the 36 apples represented in the histogram?

A) 160 grams

B) 180 grams

C) 200 grams

D) 220 grams

The graph above shows the amount of sugar contained in four drinks, A, B, C and D, as percentage of their total weights. The weight of per drink is equal. The price of the drinks A, B, C and D are $4.00, $6.00, $1.50 and $6.00, respectively. Which of the four drinks supplies the most sugar per dollar?

A) A

B) B

C) C

D) D

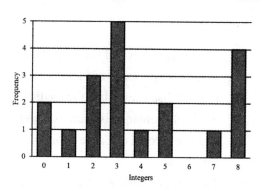

The graph above shows the frequency distribution of a list of randomly generated integers between 0 and 10. Which of the following is closest to the mean of the list of numbers?

A) 2.1

B) 2.5

C) 3.7

D) 3.9

Mobile Phone Lost

Lost times	Number of students
0	0
1	15
2	20
4	3
6	2

A researcher did a survey about how many times each of the 40 students in a class has lost their mobile phones. What is the median number of times a student lost their phones?

A) 1

B) 2

C) 4

D) 0

1

The table below shows the running distance, in miles, of Michael and Carrie for five days in a week.

Running Distances

Day	Michael	Carrie
Monday	2.03	1.52
Tuesday	1.90	1.77
Wednesday	2.51	2.12
Thursday	1.88	1.99
Friday	2.33	2.57

Based on the information in the table, which of these statements is true about the ranges and medians of the distances?

A) Both the range and median of Michael's running distances are less than the range and median of Carrie's running distances.

B) Both the range and median of Michael's running distances are greater than the range and median of Carrie's running distances.

C) The range of Michael's running distances is less than the range of Carrie's running distances, while the median of Michael's running distances is greater than the median of Carrie's running distances.

D) The range of Michael's running distances is greater than the range of Carrie's running distances, while the median of Michael's running distances is less than the median of Carrie's running distances.

2

The height, in inches, for 20 students in a class were reported, and the mean, median, range, and standard deviation for the data were found. The student with the lowest reported height was found to actually 3 inches less than her reported height. What value remains unchanged if the four values are reported using the corrected data?

A) Mean

B) Range

C) Median

D) Standard Deviation

According to a study done by Bureau of Labor Statistics, on a typical day between the years 2003 and 2006, about 16% of Americans aged 15 and older exercises. The pie chart below shows the distribution of the length of the time those people spent exercising each day.

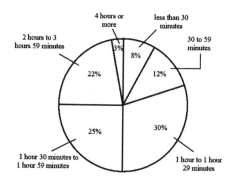

Based on the data shown, which of the following could be the median length of time spent exercising each day for those people who exercised?

A) 0 minute
B) 45 minute
C) 1 hour 15 minute
D) 2 hours 10 minute

Questions 4–5 refer to the following information.

For one week, the manager of New Oriental gathered customer feedback by randomly selecting 50 customers and asking them to complete a survey question. A total of 700 customers bought lessons at New Oriental during the week in which the survey was completed. The survey question and the survey results are shown below.

To what extent do you agree or disagree with the following statement: I would recommend this lesson to others. (Select only one of the five choices)

Survey Question Result

Response	Number of Customers
Strong agree	10
Agree	25
Neither agree nor disagree	10
Disagree	3
Strongly disagree	2
Total	50

The manager assigns the values shown below to each response choice to calculate a mean rating for the survey question.

Response	Rating
Strong agree	2
Agree	1
Neither agree nor disagree	0
Disagree	−1
Strongly disagree	−2

What is the mean rating for all 50 responses?

A) 0.27

B) 0.53

C) 0.76

D) 0.90

Which of the following is an appropriate conclusion about the customers surveyed?

A) The majority of the customers would recommend the lesson to others.

B) The majority of the customers would not recommend the lesson to others.

C) The majority of the customers would either recommend nor not recommend the lesson to others.

D) The number of the customers who would recommend the lesson to others is the same as the number of the customers who would not recommend the company to others.

Name	Rating
Alex	10min 20sec
Serene	5min 2sec
Joel	5min 51sec
John	4min 59sec
Kyle	5min 23sec

Five students ran a mile, and their times are shown in the table above. If Alex's time is removed from the data set, which of the following measures will change the least?

A) Mean

B) Median

C) Maximum

D) Range

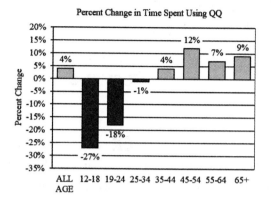

Percent Change in Time Spent Using QQ

The bar graph above shows the percent change in the average number of hours people in different age groups spent using QQ from October 2010 to October 2011. Which of the following conclusions is most appropriate based on the bar graph?

A) The majority of people who use QQ are between 45 and 64 years old.

B) Younger people spent less time using QQ in October 2011 than in October 2010.

C) The group from 12 to 18 years old spent the least amount of time using QQ over this period of time.

D) About 32% of the population spent more time using QQ in October 2011 than in October 2010.

15. Sample Survey

LEVEL 1

1

A survey was given to a random sample of 1,000 comic movie fans in Shanghai. The results of this survey should be representative of which the following populations?

A) All the comic movie fans in Shanghai

B) All comic movie fans in China

C) All movie fans in Shanghai

D) All movie fans in China

LEVEL 2

1

A survey is being conducted to determine whether residents of Shanghai are in favor of a proposal to spend $3,000,000 of local taxes on the construction of 30 additional professional football stadiums. Lenny surveyed 90 people who came to the stadium to play football. He found that 45 of those surveyed are in favor of the proposal. Which of the following statements must be true?

A) When the actual vote is taken, 50 percent of the votes will be in favor of the proposal.

B) No prediction should be made about the vote on the proposal because the sample size is too small.

C) The sampling method is flawed and may produce biased results.

D) The sampling method is not flawed and is likely to produce unbiased results.

To determine whether Peony should be the national flower of China, Amy surveyed 100 adults visiting a Peony exhibition and she found that 87% of those people vote in favor of the proposal. Which of the following statements must be true?

A) In China, 87% of people would vote for Peony to be the national flower.

B) No conclusion can be made because the sample size is too small.

C) The sampling method is flawed and may produce biased results.

D) The sampling method is not flawed and is likely to produce unbiased results.

In a study, 83 students of a school were randomly selected and approximately 78% of them reported that they stayed late almost every day. If the reported percentage is used to estimate the proportion of all the students in the school who stay late almost every day, the margin error is 21%. Which of the statement is appropriate based on the data provided?

A) Between 57% and 99% of the students in the school stay late almost every day.

B) Less than 21% of the students in the school who stay late almost every day.

C) Between 0 and 21% percent of the students in the school who stay late almost every day.

D) Approximately 22% of the surveyed students misstated how often they stay late.

A nutritionist wants to determine whether a new regimen improves the overall health of the seniors in the community. To test whether the regimen improves health, the nutritionist arranges for all the male seniors to use it for one year. The trainer will then compare the health condition data with those in last year. Which of the following will NOT improve the quality of the study?

A) Randomly assigning half of the seniors to use the new regimen while the other half uses the original ones.

B) Including members of all the male seniors from a nearby community.

C) Including those elderly who do not live in the community in the study.

D) Including members of the female seniors in the community in the study.

LEVEL 3

1

A researcher in British conducted an experiment to determine if the length of holiday affects the productivity of workers. Workers from Tom's factory were randomly selected and randomly assigned to take holiday with different length: Some with 10 days and others with 20 days. The researchers concluded that the mean productivity of workers with 20 days off is significantly higher than those with 10 days off. Based on study, which of the following statement is correct?

A) Longer holidays are likely to improve productivity for workers in Tom's factory.

B) Longer holidays are likely to improve productivity for workers, but this conclusion can not be generalized in Tom's factory.

C) It is not reasonable to conclude that the length of the holiday was the cause of difference in mean productivity for these workers.

D) It is not possible to draw any conclusion from this experiment.

2

Near the end of a US cable news show, the host invited viewers to respond to a poll on the show's website that asked, "Do you support the new federal policy discussed during the show?" At the end of the show, the host reported that 32% responded "Yes," and 65% responded "No." which of the following best explains why the results are unlikely to represent the sentiments of the population of the United States?

A) The percentages do not add up to 100%, so any possible conclusions from the poll are invalid.

B) Those who responded to the poll were not a random sample of the population of the United States.

C) There were not 50% "Yes" responses and 50% "No" responses.

D) The show did not allow viewers enough time to respond to the poll.

3

Ms. Miller is the music teachers at Lindon High school, which has 1,125students. She selected 40 of her music students at random and asked each whether they play an instrument. Of the 40 students surveyed, 31 play an instrument. Based on the design of the study, which of the following is the largest group of students to whom the results of Ms. Miller's survey can be generalized?

A) All of the Ms. Miller's music students at Lindon High School

B) All of the music students at Lindon High School

C) All of the students at Lindon High School

D) All of the students in the state

16. Polynomials

LEVEL 1

1

Which of the following is equivalent to $(x - y)(\frac{y}{x})$?

A) $y - \dfrac{y}{x}$

B) $xy - \dfrac{y}{x}$

C) $y - \dfrac{y^2}{x}$

D) $\dfrac{y}{x^2} - \dfrac{y^2}{x}$

2

If x is not equal to zero, what is the value of $\dfrac{9(2x)^2}{(3x)^2}$?

3

Which of the following is an equivalent form of the expression $9x + 21ax$?

A) $30ax^2$

B) $30(a + 2x)$

C) $(3 + 7a)x$

D) $(9 + 21a)x$

4

$$(xy^2 + 6x^2 - 2x^2y) - (-7x^2y + xy^2 - 4x^2)$$

Which of the following is equivalent to the expression above?

A) $10x^2 + 5x^2y$

B) $8xy^2 + 10x^2 - 3x^2y$

C) $2x^2 + 5x^2y$

D) $2xy^2 + 10x^2 + 5x^2y$

5

Which of the following is equivalent to $2(x^2 - x) + 3(x - x^2)$?

A) $5x^2 - 5x$

B) $5x^2 + 5x$

C) $-x^2 + x$

D) $-x^2 - x$

101

LEVEL 2

1

Which of the following is equivalent to the expression $2(x+3)(x-2)+4(x+1)(x-5)$?

A) $6x^2+14x-30$

B) $6x^2-18x-32$

C) $6x^2-14x-32$

D) $6x^2+18x-30$

LEVEL 3

1

$(ax+2)(3x^2-bx+3)=15x^3+16x^2+19x+6$

The equation above is true for all x, where a and b are constants. What is the value of ab?

A) -10

B) 10

C) -20

D) 20

2

$$(3x+3)(ax-1)-x^2+3$$

In the expression above, a is a constant. If the expression is equivalent to bx, where b is a constant, what is the value of b?

A) -4

B) -2

C) 2

D) 4

17. Radicals and Rational Exponents

LEVEL 1

1

Which of the following is equivalent to $x^{\frac{5}{3}}$, where $x > 0$?

A) $\dfrac{x^5}{x^3}$

B) $x^5 - x^3$

C) $\sqrt[3]{x^5}$

D) $\left(\sqrt[5]{x}\right)^3$

2

For a real number, $x^{11} = 27$, what is the value of x^{22}?

A) 81

B) 243

C) 729

D) 9

3

Which of the following is an equivalent form of $\sqrt[5]{a^{\frac{9}{2}}b^3}$, where $a > 0$?

A) $a^{\frac{9}{5}}b^{\frac{3}{5}}$

B) $a^{\frac{9}{5}}b^{\frac{9}{5}}$

C) $a^{\frac{9}{10}}b^{\frac{3}{5}}$

D) $a^{\frac{3}{5}}b^{\frac{3}{5}}$

LEVEL 2

1

Which of the following is equivalent to $h^{\frac{2}{9}}$?

A) $\sqrt[2]{h^9}$

B) $\sqrt[9]{h^2}$

C) $h\sqrt{h}$

D) $h\sqrt[9]{h^2}$

2

Which of the following is equivalent form of $\sqrt[5]{a^{10}b^7}$?

A) $a^2 b^{\frac{5}{7}}$

B) $a^{50} b^{\frac{5}{7}}$

C) $a^{50} b^{\frac{7}{5}}$

D) $a^2 b^{\frac{7}{5}}$

3

$$x - 3 = \sqrt{2x - 3}$$

Which is the solution set for x to the equation above?

A) {2, 6}

B) {2}

C) {6}

D) {0, 2, 6}

4

$$x + 2 = \sqrt{x + 14}$$

Which of the following values of x is a solution to the equation above?

5

If $x = 2\sqrt{5}$, and $4x = \sqrt{5y}$, what's the value of y?

A) 64

B) 2

C) 8

D) 16

6

$$(a^3 b^4)^{\frac{1}{3}}(a^3 b^4)^{\frac{1}{4}} = a^{\frac{t}{4}}b^{\frac{t}{3}}$$

If the equation above, where t is a constant, is true for all positive values of a and b, what is the value of t?

A) 4

B) 5

C) 7

D) 8

7

$$\sqrt{8x} = 6 - x$$

What value of x satisfies the equation above?

A) 2

B) 8

C) 18

D) −2

8

If $a > 0$, which of the following is equivalent to $\dfrac{2a^3}{\sqrt{3}\,a^4}$?

A) $\dfrac{2\sqrt{3}}{3}a$

B) $\dfrac{2}{3}a^{-1}$

C) $\dfrac{2\sqrt{3}}{3}a^{-1}$

D) $\dfrac{2}{3}a$

LEVEL 3

1

For a positive real number x, where $x^7 = 3$, what is the value of x^{21}?

A) $\sqrt[3]{21}$

B) 9

C) 21

D) 27

3

What is the set of all solutions to the equation $\sqrt{x + 6} = -x$?

A) $\{-2, 3\}$

B) $\{-2\}$

C) $\{3\}$

D) There is no solution to the given equation

2

If $125^x = 5$ and $12^{x+y} = 144$, what is the value of y?

A) −1

B) $\dfrac{1}{3}$

C) $\dfrac{5}{3}$

D) $\dfrac{7}{3}$

4

$$\sqrt{9x^2} - 3x = 0$$

Which of the following values of x is NOT a solution to the equation above?

A) −4

B) 0

C) 1

D) 3

18. Plane Geometry

LEVEL 1

1

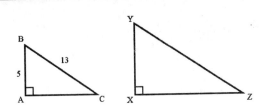

Right triangles *ABC* and *XYZ* above are similar. What is the value of tan *Z*?

A) $\frac{5}{13}$

B) $\frac{5}{12}$

C) $\frac{12}{5}$

D) $\frac{13}{5}$

2

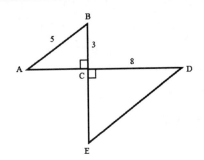

Note: Figure not drawn to scale

In the figure above, $\angle B$ is congruent to $\angle E$. What is the value of sin *D*?

A) $\frac{3}{5}$

B) $\frac{4}{5}$

C) $\frac{8}{5}$

D) $\frac{4}{3}$

3

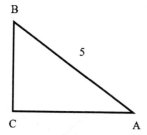

In right triangle *ABC* above, the tangent of *B* is $\frac{4}{3}$, what is the length of $\overline{AC}$?

A) 4

B) 5

C) 3

D) 10

4

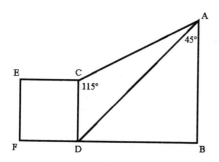

In the figure above, *AB* and *CD* are parallel, and *EFDC* is a square. If the measure of $\angle ACD = 115°$, $\angle DAB = 45°$, what is the measure of $\angle CAD$?

A) 20°

B) 30°

C) 40°

D) 50°

5

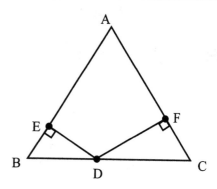

Triangle *ABC* above is isosceles with *AB = AC* and *BC* = 36. The ratio of *DE* to *DF* is 5:7. What is the length of $\overline{DC}$?

A) 9

B) 15

C) 18

D) 21

6

Each angle of $\triangle ABC$ is congruent to one of the angles of $\triangle QRS$. If *AB* = 3, *BC* = 4, *AC* = 6 and *RS* = 12, what is one possible value for the perimeter of $\triangle QRS$?

LEVEL 2

1

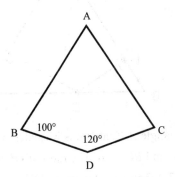

In the figure above, $\overline{AB} = \overline{AC}$ and $\overline{BD} = \overline{DC}$. What is the value of $\angle A$ in degrees? (Disregard the degree sign when gridding your answer.)

2

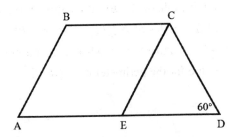

In the figure above, $\overline{BC}$ and $\overline{AD}$ are parallel, $\overline{AB}$ and $\overline{EC}$ are parallel, $CD = CE$, and the measure of $\angle CDE$ is 60°, what is the measure of $\angle ABC$?

A) 90°

B) 120°

C) 150°

D) 160°

3

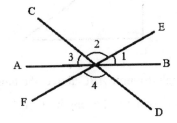

In the figure above, $\angle 1 = 30°$ and $\angle 4 = 110°$, what is the value of $\angle 3$?

A) 30°

B) 40°

C) 50°

D) 60°

4

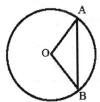

If the radius above is x, $\angle AOB = 120°$, and O is the center of the circle, what is the length of chord AB in terms of x?

A) $\sqrt{2}x$

B) $\sqrt{3}x$

C) $\dfrac{x}{\sqrt{2}}$

D) $\dfrac{x}{\sqrt{3}}$

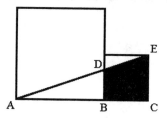

The figure above shows that the shaded triangular region with AB = 9cm, and BD = 3cm. Area of the bigger square is 4 times the area of the little one. What's the area of the trapezoid $BDEC$?

A) 16.875

B) 32

C) 30.5

D) 30.125

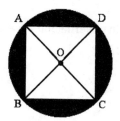

The diagram above shows a sign where the four shaded area will be removed from the circle. The radius is 5cm. What is the sum of the shaded areas?

A) 25

B) $25\pi - 25$

C) $25\pi - 50$

D) 10

A child is playing a piece of string and form it into a shape of square firstly and into a shape of regular pentagon. If each side of the square is 2.5 inches longer than each side of the pentagon, how long, in inches, is the string?

A) 10

B) 30

C) 40

D) 50

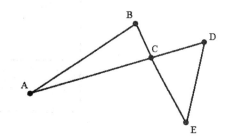

Note: Figure not drawn to scale

In the figure above, $\angle BAC = 20°$, $\angle CED = 35°$, $\angle ABC = 80°$, what is the angle measure of $\angle CDE$, in degree?

A) 55

B) 65

C) 75

D) 85

The side length of Square *A* is 4 times as long as side length of Square *B*. How many times larger is the area of Square *A* than the area of Square *B*?

A) 3

B) 4

C) 9

D) 15

LEVEL 3

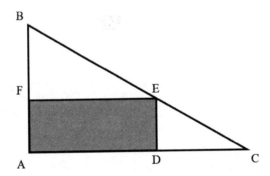

In the figure above, *ABC* is a right triangle and $2AC = 3AB$. In quadrilateral $FE = 2ED$, the area of the shaded region is what fraction of the area of triangle *ABC*?

A) $\dfrac{13}{20}$

B) $\dfrac{12}{25}$

C) $\dfrac{24}{49}$

D) $\dfrac{49}{24}$

2

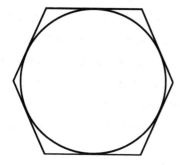

The length of each side of the hexagon above is 6. A circle is inscribed in the hexagon. What is the area of this circle?

A) 9π

B) 18π

C) 27π

D) 36π

Questions 3–4 refer to the following information.

An instrument on a bike shows the number of revolutions per minute made by each tire of bike. In each revolution, the bike travels a distance equal to the circumference of one of its tires. The circumference of each tire is equal to $2\pi r$, where r is radius of the tire.

3

Amy rode bike for 5 minutes all the way from school to her home in a street along a straight line. The radius of tire on Amy's bike is 0.5 meter. To the nearest hundredth, what is the distance, in kilometer, from her school to her home, when the instrument is showing 100 revolutions per minute?
(1 kilometer = 1000 meters)

4

Amy need to repair her bike and need to increase the radius of his bike. The original radius is 0.5 meter, and radius has increased 16% after repair. What is the circumference of new tire, to the nearest meter?

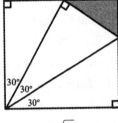

$5\sqrt{3}$

The figure above shows that the shaded triangular region has been removed from a rectangular tile with a length of $5\sqrt{3}$ cm. Of the following. Which best approximates the area, in square centimeters, of the removed shaded triangle?

A) 5.4

B) 11.4

C) 23.4

D) 50.4

$$S = 2Cr$$

The formula above can be used to approximate the surface area S of a planet using its average radius r and average circumference C. The surface area of Saturn is 4.6×10^{10} square kilometers. Of the following, which best approximates the average radius, in kilometers, of Saturn?

A) 1,568

B) 60,500

C) 273,300

D) 704,300

Carey rode a bike to a theater to watch the Mozart L'opera Rock. He rode 4,500 m east and then 2,300 m north and then 300 m west and he finally arrived at the theater. If there is a straight level path from his home to the theatre, what would be the length, rounded to the nearest meter, of the path?

A) 4,788

B) 7,100

C) 5,053

D) 6,500

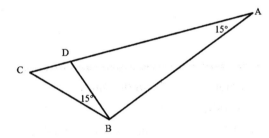

In the figure above, which of the following ratios has the same value as $\dfrac{AB}{AC}$?

A) $\dfrac{BD}{BC}$

B) $\dfrac{BC}{AC}$

C) $\dfrac{BD}{DC}$

D) $\dfrac{DC}{BC}$

19. Solid Geometry

LEVEL 2

1

$$S = 2Cr$$

The formula above can be used to approximate the surface area S of a planet using its average radius r and average circumference C. The volume V of a planet can be expressed in terms of its surface area S and its average radius r by formula $V = \frac{1}{3}Sr$, Which of the following expresses the planet's average radius, r, in terms of its volume and its average circumference?

A) $r = \frac{2V}{3C}$

B) $r = \frac{3V}{2C}$

C) $r = \sqrt{\frac{2V}{3C}}$

D) $r = \sqrt{\frac{3V}{2C}}$

2

The shape of Metrocity in Zikawei looks like a ball and the radius is about 1,300 m. To the nearest cubic meters, what is the maximum amount of space that the Metrocity can hold?

A) 9.2×10^9

B) 9.2×10^{10}

C) 4.6×10^9

D) 4.6×10^{10}

3

What is the height, in centimeters, of a rectangular prism that has a length of 4 centimeters, a width of 9 centimeters, given the volume of the prism 360 cubic centimeters?

LEVEL 3

1

A right circular cylinder has a height of 8 inches and is $\frac{3}{8}$ full of water. If the amount of water in the can is 27π cubic inches, what is the diameter, in inches, of the can?

20. Coordinate Geometry

LEVEL 1

1

$$(x - 33)^2 + (y - 21)^2 = 11$$

The graph of the equation above is a circle in the *xy*-plane. What is the area of the circle?

A) 55π

B) 12π

C) 11π

D) 5π

2

A circle is graphed in the *xy*-plane. If the circle has a radius of 7 and the center of the circle is at (–1, 3), which of the following could be an equation of the circle?

A) $(x - 1)^2 + (y + 3)^2 = 7$

B) $(x - 1)^2 - (y + 3)^2 = 7$

C) $(x + 1)^2 + (y - 3)^2 = 49$

D) $(x + 1)^2 - (y - 3)^2 = 49$

LEVEL 2

1

The graph of $x^2 + 8x + y^2 - 6x - 30 = 0$ in the xy-plane is a circle. What is the radius of the circle?

A) $\sqrt{30}$

B) $\sqrt{5}$

C) $\sqrt{55}$

D) $\sqrt{14}$

2

The graph $x^2 + 6x - 13 + y^2 + 4y + 2 = 0$ in the xy-plane is a circle. Which is the diameter of the circle?

A) 5

B) 8

C) $\sqrt{23}$

D) $4\sqrt{6}$

3

Which of the following equations describes a circle with a radius 5 and has a center of (2, 3) when graphed in the xy-plane?

A) $x^2 + 2x + 3y + y^2 = 11$

B) $x^2 - 4x - 6y + y^2 = 21$

C) $x^2 - 4x - 6y + y^2 = 12$

D) $x^2 + 4x + 6y + y^2 = 11$

21. Trigonometry

LEVEL 1

1

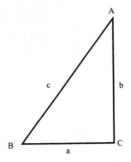

Given the right triangle ABC above, which of the following is equal to $\frac{a}{c}$?

A) $\sin B$ B) $\cos B$ C) $\tan B$ D) $\cos A$

LEVEL 2

1

If $\cos x° = a$, which of the following must be true for all value of x?

A) $\sin x° = a$ B) $\cos (90° - x°) = a$ C) $\sin (90° - x°) = a$ D) $\cos (x^2)° = a^2$

LEVEL 3

1

Triangle ABC has right B. If $\sin A = \frac{4}{5}$, what is the value of $\tan C$?

22. Degree and Radian

LEVEL 1

1

An angle with a measure of $\frac{4\pi}{5}$ radians has a measure of k degrees, where $0 < K < 360$, what is the value of k?

LEVEL 2

1

In the circle below, radius $\overline{OA}$ has length 2. The measure of $\angle AOB = \angle BOC = \angle COD$, and the length of arc $ABCD$ is 2π. What is the measure, in degree, of $\angle COD$? (Disregard the degree symbol when gridding your answer.)

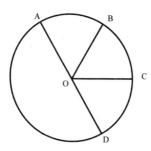

2

An angle with a measure of $\frac{11\pi}{6}$ radians has a measure of d degrees, where $0 \le d \le 360$. What is the value of d?

23. Complex Number

1

What is the value of $(2 + 8i)(1 - 4i) -$ $(3 - 2i)(6 + 4i)$? (note: $i = \sqrt{-1}$)

1

$$i^{2018} + i^{2019} + i^{2020} + i^{2021}$$

The complex number expression above can be rewritten in the form of $a + bi$, where a and b are real numbers. What is the value of $|a| + |b|$? (note: $i = \sqrt{-1}$)

118

24. Others

LEVEL 2

1

If m and n are negative integers, which of the following is NOT equivalent to $\frac{n}{m}$?

A) $\dfrac{\frac{1}{-m}}{-n}$

B) $\dfrac{-(-n)}{-m}$

C) $\dfrac{-n}{-m}$

D) $\dfrac{1}{\frac{m}{n}}$

Questions 2–3 refer to the following information.

Erythrocyte content ($10^{12}/L$)

	Male	Female
0-1 year	2	1.7
2-7 years	2.5	2.3
8-12 years	3.1	3.5
13-18 years	4.2	3.4
18 years+	5.1	3.6

2

The table shows the number of Erythrocyte content in 10^{12} per liter. John and Anya are 27 and 25 years old respectively. What is the total Erythrocyte content of them, in 10^{12} per liter?

3

Echo is 15 years old and she has a cousin David who is 11 years old. How much Erythrocyte content does she have more than her cousin in 10^{12} per liter?

As a car moves along a straight line, it speeds up quickly at first and then after 5 seconds it moves at a constant velocity of 10m/s for 20s and later on it decelerate uniformly to 10m/s for 10s. Which of the graph could represent the velocity of the car over time in s?

A)

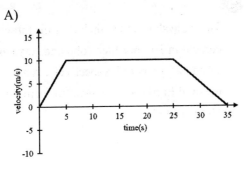

B)

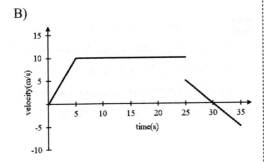

C)

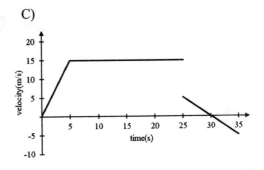

D)

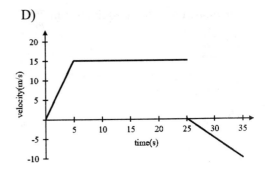

FINAL EXAM 1

Math Test – No Calculator
25 MINUTES, 20 QUESTIONS

1

Temperature in °C	Physical state of Aluminum
1200	Solid
1300	Solid
1400	Solid
1500	Liquid
1600	Liquid

The table above shows the physical state of aluminum at various temperatures T, measured in degrees Celsius (°C). Which of the following inequalities could describe the set of temperatures for which aluminum is in a solid state?

A) $T \leq 1200$

B) $T \leq 1400$

C) $T \geq 1550$

D) $T \geq 1650$

2

Which of the following is equivalent to the expression $4x(x^2 + 3) - 5(x^2 + 4)$?

A) $4x^5 - 5x^4 + 36x - 20$

B) $16x^4 + 60x^3 - 20x^2 - 75x$

C) $4x^3 - 5x^2 + 12x - 20$

D) $4x^3 - 5x^2 + 4x + 20$

3

$$10x - 12 = 48$$

If x is the solution of the equation above, what is the value of $3x - 5$?

A) 4

B) 18

C) 13

D) 24

4

If $(3c + 6d)^2 = 144$. which of the following is a possible value of $c + 2d$?

A) 5

B) 6

C) 8

D) 4

5

Jacob bought some pants and t-shirts. The pants cost \$90 per piece, and the t-shirts cost \$130 per piece. If Jacob spent \$920 in total and bought 3 pants, how many t-shirts did Dan buy?

A) 4

B) 5

C) 6

D) 7

6

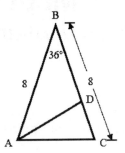

For isosceles triangle ABC shown above, $AB = BC = 8$ and the measure of angle ABC is 36 degrees. If $\angle BAC$ is bisected by AD, which of the following statements must be true?

A) $AB = AC = BC$

B) $AD = BD = AC$

C) $BD = CD = AC$

D) $AB = BD = AD$

7

$$2x^2 - 20x - 9 = 2$$

The solutions of the equation above are t and u. what is the value of $t + u$?

A) 4

B) 6

C) 8

D) 10

If Daniel rides a bike at a constant speed of 12 miles every 4 hours, which of the following functions represents the number of miles, m, Daniel rides in t hours?

A) $m(t) = 12t$

B) $m(t) = 3t$

C) $m(t) = 6t$

D) $m(t) = 4t$

A family auto repair company prepares c cups of punch for a party of n people, where $c = 5n + 8$. According to the equation, how many additional cups of punch does the manager prepare for each additional person at the party?

A) 1

B) 3

C) 5

D) 8

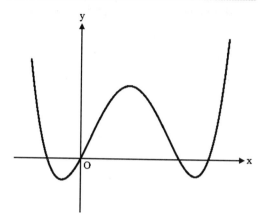

Which of the following could be an equation of the graph above?

A) $y = x(x - 1.1)(x - 3)(x - 5)$

B) $y = x(x - 39)(x - 30)(x + 10)$

C) $y = x(x - 70)(x + 20)(x + 25)$

D) $y = (x - 500)(x - 100)(x + 175)(x + 150)$

$$\left(4x^{\frac{5}{2}}y^{\frac{1}{3}}\right)^{3}$$

Which of the following expressions is equivalent to the one above?

A) $4x^{\frac{15}{2}}y$

B) $64x^{\frac{15}{2}}y$

C) $4x^{\frac{15}{8}}y^{\frac{1}{27}}$

D) $64x^{\frac{1}{2}}y^{\frac{1}{3}}$

Bobby has a savings account where he made an initial deposit of m dollars and had no deposits or withdrawals since then. The amount of money, P, in the account t years after the initial deposit is given by the equation below.

$$P = m(1.02)^t$$

By what percent did the amount of money in the account grow from the beginning of year 2 to the beginning of year 4?

A) 1.04%

B) 1.20%

C) 2.01%

D) 4.04%

The organizer of an event, Joy can spend up to $1,830 on prizes. Each prize costs either $50 or $100, and she must purchase a minimum of 20 prizes. What is the maximum number of $100 prizes she could purchase?

A) 12

B) 16

C) 20

D) 32

$$y - 4 = \frac{1}{4}(x - 1)$$

$$y - 4 = \frac{1}{4}(x - 1)^2$$

One solution, (x, y), of this system of equations is $(1, 4)$. What is the y-value of the other solution?

A) 2

B) 3

C) $\frac{22}{4}$

D) $\frac{17}{4}$

$$\sqrt{x + 4} = \frac{1}{4}x + 2$$

What is the sum of the solutions of the equations above?

A) 4

B) 2

C) 0

D) −2

16

Tracy and Sally will share a total of 240 toy cars. Tracy will receive half as much as Sally. What amount of toy cars, will Sally receive?

17

If $f(x) = 5x^2 - 4x + 8$, what is the y-intercept of the graph of $g(x) = 4f(x)$ in the xy-plane?

18

If $x + 2y = 250$ and $3x - 4y = 500$, what is the value of y?

19

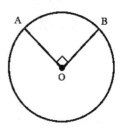

In the circle above, O is the center and $OB = 5$. If the length of arc AB is $a\pi$. where a is a constant, what is the value of a?

20

If $\sin 48° = \cos x°$, where $0 < x < 180$, what is a possible value of x?

Math Test – Calculator

55 MINUTES, 38 QUESTIONS

1

Johnson has read 150 pages of a novel that contains 330 pages. If Johnson continues to read at a rate of 15 pages per day, how many days will it take him to finish the book?

A) 9

B) 12

C) 18

D) 22

2

If $a = 9$, how much greater than $2a - 1$ is $6a - 5$?

A) 32

B) 22

C) 24

D) 28

3

Ramsy had x dollars and bought y stamps that costs $0.6 each. Which of the following expressions represents the amount of money, in dollars, Ramsy had left after he bought the stamps?

A) $x - 0.6y$

B) $x + 0.6y$

C) $100x - 6y$

D) $100x + 6y$

4

Lucus can fill an empty tank with N18 petro for $8.75 when petro costs $1.75 per gallon. How much will it cost Lucus to fill the same empty tank with N20 petro that costs $2.00 per gallon?

A) $ 9.00

B) $10.00

C) $11.00

D) $12.00

5

Eric works in a yoghurt shop where he is paid at a same hourly rate each day. He was paid $71.25 for working 7.5 hours on Monday. If he worked 8 hours on Tuesday, how much was he paid on Tuesday?

A) $51.00

B) $57.00

C) $76.00

D) $79.00

6

If 11 plus y is equal to 6, what is the value of y?

A) −4

B) −5

C) 5

D) 4

7

Ryan's bank charges 2% interest on his credit card balance each month. The balance on his credit card last month was $1,400. How much interest did the bank charge on the balance that month?

A) $0.28

B) $2.80

C) $14.00

D) $28.00

8

To determine if age and gender are related to pet ownership at his school, Mark selected a random sample of 70 male students who are 18 to 19 years old from the school and a random sample of 80 female students who are 19 to 20 years old from the school. For each student, he recorded the student's age, gender, and whether the student owned a pet. Which of the following provides the best explanation for why Mark cannot draw a valid conclusion from this study?

A) The sample sizes are too small.

B) The two samples are not of equal size.

C) Mark will be unable to tell whether a difference in pet ownership is related to age because the two age groups are too close in age.

D) Mark will be unable to tell whether a difference in pet ownership is related to gender because of the difference in age. Similarly, he will be unable to tell whether a difference in pet ownership related to age.

Which of the following ordered pairs (x, y) satisfies both of the equations $y = x^2 - 7x + 10$ and $y = -x + 5$?

A) $(-1, 8)$

B) $(0, 11)$

C) $(1, 4)$

D) $(5, -4)$

Which of the following expressions is equivalent to $\dfrac{x^2 + 2x - 15}{x^2 - 9} \times \dfrac{x + 3}{x}$?

A) $\dfrac{x - 5}{x}$

B) $\dfrac{x + 5}{x}$

C) $\dfrac{x^2 - 5x}{x^2 - 6x + 9}$

D) $\dfrac{x^2 - 2x - 15}{x^2 - 3x}$

Rocks are collected by students and they classify a rock collection according to the predominant color of each rock and type of rock. The number of rocks in each classification as shown below.

Rock color	iqneous	metamotphic	sendimentary	Total
Black	16	3	1	20
Brown	8	11	45	64
Gray	14	0	38	52
Red	22	0	6	28
Pink	9	0	0	9
Tan	0	7	1	8
White	4	14	19	37
Total	73	35	110	218

A student selects a random rock and notices it is tan. Given this information, which of the following is closest to the probability that the selected rock is sedimentary?

A) 0.17   B) 0.125   C) 0.12   D) 0.15

12

A total of 300 surgeries, with a combined duration of 1,130 hours, were required for an intern to sign an official contract. In which of the following equations does x represent the average (arithmetic mean) duration, in hours, of the 300 surgeries?

A) $1,130 = 300x$
B) $300 = 1,130x$
C) $x = 300 \times 1,130$
D) $x = 300 + 1,130$

13

A circle in the xy-plane has center $(-6, -6)$, and the point with coordinates $(-3, -10)$ is on the circle. What is the <u>diameter</u> of the circle?

A) 5
B) 10
C) 15
D) 25

14

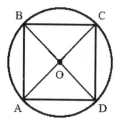

In the figure above, square $ABCD$ is inscribed in the circle with center O and radius 5. What is the length of AB?

A) 10
B) $5\sqrt{2}$
C) $10\sqrt{2}$
D) $5\sqrt{3}$

15

Cheeto advertises that their bags of chips contain, on average. 1 pound of pretzels. To test this, Sam selected at random and weighed the contents of each bag. Based on his measurements, Sam estimated that the average weight of a bag of chips produced by the company is 0.96 pounds, with a margin of error of 0.13 pounds. Which of the following is the most plausible conclusion about the true average weight w, in pounds, of a bag of pretzels produced by the company?

A) $w = 0.96$
B) $0.83 \leq w \leq 0.96$
C) $0.83 \leq w \leq 1.09$
D) $0.96 \leq w \leq 1.09$

Questions 16–18 refer to the following information.

Estimated Total Daily Water Usage, in Millions of Gallons, for Shanghai in 2005

Type	Ground	Surface
Fresh	592	1,000
Saline	0.01	5,600

The table above shows the estimated water usage, in millions of gallons per day, by source and type, for Shanghai in 2005. In 2005, Shanghai had an estimated population of 8,720,000.

16

Which of the following is closest to the proportion of estimated daily usage of water that was fresh?

A) 0.22
B) 0.14
C) 0.18
D) 0.24

17

Approximately what percent of the estimated Daily water usage was not groundwater but fresh?

A) 8%
B) 14%
C) 20%
D) 34%

18

Which of the following best approximates the estimated water usage of groundwater in acre-feet per year? (1 million gallons = 3.07 acre-feet)

A) 0.05
B) 1.19
C) 1817.4
D) 6×10^5

19

The value of x is 3 more than the value of y. The sum of $3x$ and $4y$ is 20 less than the value of $8x$. What is the value of x?

A) 5
B) 6
C) 8
D) 12

Questions 20–21 refer to the following information.

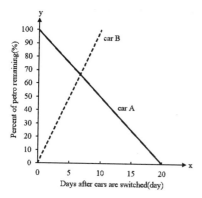

Days after cars are switched(day)

Jerry has two cars. When car B is out of petro, he parks it at the garage and refills it up. He drives car A during that period. The graph above relates the percent of petro remaining for the two cars to the time, in days, after they are switched.

20

Which of the following gives the percent of petro remaining, y, for car A in terms of the number of days, x, after car A and car B are switched?

A) $y = -5x + 20$

B) $y = -5x + 100$

C) $y = -10x + 100$

D) $y = -20x + 100$

21

Which of the following can be concluded from comparing the line representing car A to the line representing car B?

A) The rate at which car A uses petro is greater than the rate at which car B burns gas.

B) The rate at which car A refills is greater than the rate at which car B burns gas.

C) The rate at which car B refills is greater than the rate at which car A burns gas

D) The rate at which car B burns gas is greater, than the rate at which car A refills.

22

In physics, Hooke's law states that the length change of a spring is $F = 1.827x$, in which F represents the force and x represents the length change of the spring. Which of the following statement is true?

A) The change of length is positively related to the force applied to the spring.

B) The change of length is negatively related to the force applied to the spring.

C) There is no relationship between the change of length and the force applied to the spring.

D) The change of length would decrease if less force is applied to the spring.

$$\frac{x}{x-4} = \frac{3x}{3}$$

Which of the following represents all the possible value of x that satisfy the equation above?

A) 0 and 3

B) 0 and 5

C) 3 and 5

D) 5

Questions 24–25 refer to the following information.

$$F = \frac{mV^2}{r}$$

The formula above relates the centripetal force F, in newton, acting on an object traveling in a circular path, to the object's mass m, its velocity v and the radius r of its path.

If the velocity of the object is tripled, which of the following equations expresses the new centripetal force, N of the object, in terms of the original centripetal force F?

A) $N = 9F$

B) $N = 3F$

C) $N = \frac{1}{3}F$

D) $N = \frac{1}{2}F$

Of the following equations, which is NOT equivalent to the formula for centripetal force?

A) $Fr = mV^2$

B) $\frac{V^2}{F} = \frac{m}{r}$

C) $\frac{Fr}{m^2} = \frac{V^2}{m}$

D) $r = \frac{mV^2}{F}$

In 2015, the red county Railroad made a plan to reduce the number of railroad cars by 13% cars per year for each of the next 15 years. Which of the following types of expressions could be used to model the number of cars Red County Railroad has in service n years after 2015, where n is an integer from 1 to 15?

A) $a + bn$, where a is a positive constant and b is a negative constant

B) $a + bn$, where a is a negative constant and b is a positive constant

C) $a(b)^n$, where a is a positive constant and b is a constant such that $b > 1$

D) $a(b)^n$, where a is a positive constant and b is a constant such that $0 < b < 1$

A piece of gold is initially valued at $150. Every month the value of the piece of gold increases by 3% of its value the previous month. Which of the following represents the value $Q(t)$, in dollars, of the piece of jewelry at the end of n months?

A) $150 + 1.03n$

B) $150 + 0.03n$

C) $150(1.03)^n$

D) $150(0.03)^n$

The circular dock shown above has diameter 14 inches, and its minute hand has length 6 inches. It is placed on the wall so that the center of the clock it 66 inches above the ground. Which of the following graphs could represent the distance from the tip of the arrow of the minute hand to the ground with respect to time from 10 a.m. to 11 a.m?

A)

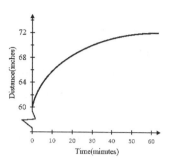

B)

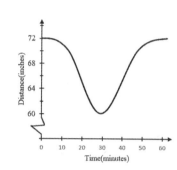

C)

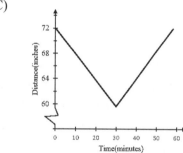

D)

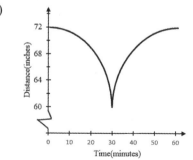

29

Pierre plans to make at least 20 pounds of a fruit mix that will consist of peaches and dragon fruit. If he wants the mix to be at least 50% peach by weight, which of the following systems of inequalities represents, p, and the number of pounds of dragon fruit, f?

A) $p+f\leq20$
   $0.5p\leq f$

B) $p+f\geq20$
   $0.5f\leq p$

C) $p+f\geq20$
   $f\leq p$

D) $p+f\leq20$
   $0.5p\geq f$

30

The speed of sound in air is 340 meters per second. Given that 1 kilometer is equal to approximately 0.62 miles, which of the following is closest to the speed of sound in a vacuum, in miles per hour?

A) 700

B) 730

C) 760

D) 790

31

A 50-foot-long stick is leaning against a building, as shown below.

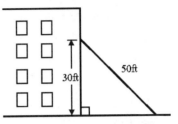

The top of the stick is resting on the budding at a point 30 feet above the ground. How many feet from the base of the budding is the bottom of the stick?

32

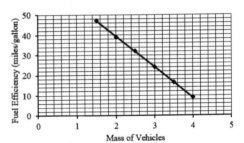

The scatterplot above shows the fuel efficiency, in miles per gallon, of a variety of vehicles weighing between 1.5 and 4 tons. Based on the line of best fit to the data, what is the expected fuel efficiency, in miles per gallon, of a vehicle that weighs 3 tons?

Professor determines a student's course grade by computing the mean of 4 scores, including 3 unit test scores and the final exam score. If a student misses a unit test, the final exam score is used once as the final exam score and once in place of the missing unit test score. What is the course grade for a student who receives scores of 76 and 80 on 2 unit tests, misses 1 unit test, and earns a score of 92 on the final exam?

Anny used a car that was originally priced at $25,000 and then had a 30% discount. If an additional 25% was taken off the reduced price, what was the price, in dollars, of the used car after the two reductions were applied? (Disregard the $ sign when gridding your answer.)

Lucas is selling a type of phone designed for elder people. He uses the inequality $100,000 \leq x \leq 300,000$ to estimate the profit x, in dollars, he could make from selling 50,000 phones. What is the maximum profit, in dollars, he could expect to make from each cell phone? (Disregard the $ sign when gridding your answer.)

The function $f(x) = 58x^2 - 200x + 1500$, where x represents the number of years after 2000, can be used to model the number of college-bound students, $f(x)$, who take a test to receive college credit for an economic course. Based on the model, how many more students took the test in 2018 than in 2015?

On May 20, 2016, Jane's tenth grandchild was born, and the average (arithmetic mean) age of his other 9 grandchildren was 11 years. What will be the average age, in years, of Jane's 10 grandchildren on May 20, 2018?

In the system of linear equations above, c is a nonzero constant. The graphs of the equations are two lines in the xy-plane intersect at $(k, -22)$. What is the value of k?
$$5y + 9x = 2c$$
$$3y + 9x = 3c$$

FINAL EXAM 2

Math Test – No Calculator

25 MINUTES, 20 QUESTIONS

1

A zoo sells a family ticket combo at the price of $30 and an additional $4 insurance for each family member. Which of the following represents the total cost c, in dollars, to visit a zoo for a family of d members?

A) $c = 34d$

B) $c = 30(d + 24)$

C) $c = 4d + 30$

D) $c = 30d + 4$

2

Jenny is baking pizzas for the Christmas party. A pizza can be shared by 5 children. If 100 children are invited to the party. How many children would not get a slice of pizza if c pizzas are baked?

A) $100 + 5c$

B) $100 - 5c$

C) $5c - 100$

D) $\dfrac{100c}{3}$

3

$$y = 4x^2, \quad y = -4x$$

Which of the following ordered pairs (x, y) is a solution to the system of equations above?

A) $(-1, -4)$

B) $(-1, 4)$

C) $(1, -4)$

D) $(1, 4)$

4

$$(3x - 5)(x^2 - 3x + 4)$$

Which of the following is equivalent to the expression above?

A) $3x^3 - 14x^2 + 27x - 20$

B) $3x^3 - x^2 + 27x - 20$

C) $3x^3 - 5x^2 + 8x - 20$

D) $3x^2 - x - 1$

There are two kinds of puzzles: 50-piece puzzle and 85-piece puzzle. If a factory can produce 1,000 pieces in total, what is the maximum number of the 50 pieces puzzles can be produced if at least 10 packs of the 85-piece puzzles should be produced?

A) 2

B) 3

C) 5

D) 8

A movie theater has two types of tickets: superior tickets that cost $20 per movie and basic tickets that cost $16 per movie. On a given movie night, the theater received $840 from selling 46 tickets. How many superior tickets were sold on that night?

A) 20

B) 21

C) 26

D) 30

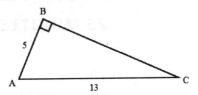

Triangle *ABC* above is similar to triangle *DEF* (not shown) where vertices *A*, *B*, and *C* correspond to vertices *D*, *E*, and *F* respectively. If *DE* = 15, what is the area of triangle *DEF*?

A) 45

B) 90

C) 150

D) 270

$$3x - 8 = y, \ 3y - 8 = x$$

If (x, y) is a solution to the system of equations above, what is the value of $x + y$?

A) 6

B) 8

C) 15

D) 10

9

A charity is gathering donation for a school in Wenchuan by selling movie tickets. For each child ticket sold, the theater will donate \$4.5. For each adult ticket sold, the theater will donate \$7. Assuming the theater will sell 150 child tickets, which inequality can be used to determine the number of adult tickets, x, that will need to sell in order to donate at least \$2,000 to the school?

A) $7x \leq 1,325$

B) $7x \geq 1,325$

C) $7x \leq 1,500$

D) $7x \geq 1,500$

10

Which of the following is equivalent to $\dfrac{x^2 y}{x^4 y^{\frac{1}{4}}}$ for all $x > 0$ and all $y > 0$?

A) $\dfrac{\sqrt[4]{y^3}}{x^2}$

B) $\dfrac{x^2}{y^{\frac{1}{2}}}$

C) $x^2\sqrt{y}$

D) $x^6\sqrt{y^3}$

11

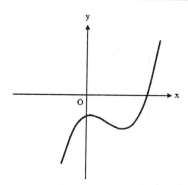

Which of the following could be the equation of the graph in the xy-plane above?

A) $y = (x^2 + 3)(5 - x)$

B) $y = (x^2 + 3)(x - 5)$

C) $y = (x^2 + 3)(x + 5)$

D) $y = (x^2 - 9)(x + 3)$

12

The kinetic energy of an object, k, in joules, can be represented by the formula $k = \dfrac{mv^2}{2}$, where m is the mass of the object, in kg, and v is the speed at which the object is traveling. In meters per second. Which of the following correctly shows the speed of the object in terms of its kinetic energy and mass?

A) $v = \sqrt{\dfrac{2k}{m}}$

B) $v = \sqrt{\dfrac{2m}{k}}$

C) $v = \dfrac{\sqrt{2m}}{k}$

D) $v = \sqrt{2mk}$

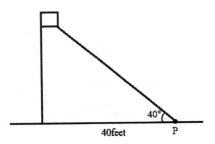

40°

40feet   P

In the figure above, P is a point on the ground 40 feet from the base of a flagpole. The angle of elevation from point P to the top of the flagpole is 40°, and the tangent of 40° is approximately 0.84. Of the following, which is the closest to the height of the flagpole?

A) 25 feet

B) 23 feet

C) 30 feet

D) 34 feet

Which of the following is equivalent to

$$\frac{2}{x+2} + \frac{1}{x+1}, x > 0?$$

A) $\dfrac{3}{x+3}$

B) $\dfrac{2x+3}{x+3}$

C) $\dfrac{2x+3}{x^2+3x+2}$

D) $\dfrac{3x+4}{x^2+3x+2}$

If $-2x + 5 = -x - 4$, what is the value of x?

A) -9

B) 9

C) -1

D) 1

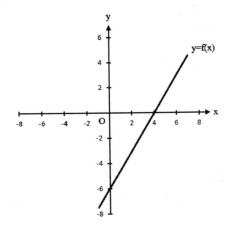

The graph of the linear function f is shown in the xy-plane above. What is the y-intercept of the line that is parallel to the graph of f and passes through the point $(-2, 0)$?

The expression $h = -5t^2 + 25t + 4$ represents the height, in meters, of a rocket t seconds after it was launched straight up into the air. What was the height of the shooting platform?

The quadratic equation $9x^2 + 125 = 140$ has two solutions. What is the sum of the solutions?

The product of the two complex numbers $5 + 3i$ and $5 + 2i$ is written in the form $a + bi$, where a and b are real numbers. What is the value of b?

$$an + 8 = a(n + 3) + 1.7$$
If $n = -5$ in the equation above, what is the value of a?

Math Test – Calculator
55 MINUTES, 38 QUESTIONS

1

Johnny has eaten m ice-creams each day for 8 days. In terms of m, what is the total number of ice-creams Johnny ate in the 8 days?

A) $8m$

B) $\dfrac{8}{m}$

C) m^8

D) $8 + m$

2

An orange juice machine can produce a maximum of 300 pounds of orange juice each day. If 1 cubic foot of juice weights about 57.2 pounds, which of the following best approximates the maximum number of cubic feet of juice the machine can produce in one day?

A) 0.52

B) 5.2

C) 52

D) 520

3

$(x + 2) - 5(x + 2) = 0$ What is the solution x to the equation above?

A) –2

B) –1

C) 0

D) 1

In a large college, of the students in English-degree program, 300 were selected at random and asked how many cups of coffee they drink on average each month. The results of this survey can be best generalized to which of the following populations?

A) All students in the same large college

B) Any sample of 300 students in the same large college

C) All students in English-degree program in any College

D) All students in English-degree program in the same large college

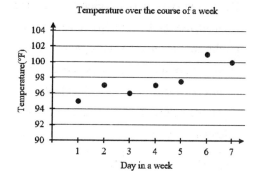

Temperature over the course of a week

The daily high temperature is recorded in the table, in degrees Fahrenheit (°F), over the course of a week in a city as shown above. Between which two consecutive days did the greatest increase in high temperature take place?

A) Day 1 and Day 2

B) Day 4 and Day 5

C) Day 5 and Day 6

D) Day 7 and Day 8

Over a two-year period, scientists searched for neutrinos (subatomic particles) that are produced outside the solar system. The energy, E, of those neutrinos is measured in units of 1,012 electron volts (TeV). The table below shows the results of the scientists' investigation:

Neutrino energy E (Tev)	Number of neutrinos detected
$E \leq 30$	1
$30 < E < 60$	10
$60 < E \leq 90$	7
$90 < E \leq 120$	3
$120 < E \leq 150$	0
$E > 150$	7

What percentage of the all detected neutrinos have more than 150 Tev neutrino energy?

A) 18%

B) 25%

C) 64%

D) 72%

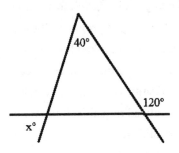

In the figure above, the sides of a triangle are extended as shown. What is the value of x?

A) 80

B) 70

C) 60

D) 40

In a basketball league, each team received 5 points for a win, 2 point for a tie, and 0 points for a loss. The Blue Jersey team has no losses and a total of 41 points after playing 18 games. Which system of equations could be used to solve for the number of wins and ties, where w is the number of wins and t is the number of ties?

A) $\begin{cases} 5w + 2t = 41 \\ w + t = 18 \end{cases}$

B) $\begin{cases} 5w + 2t = 18 \\ w + t = 51 \end{cases}$

C) $\begin{cases} 5w + 5t = 41 \\ w + t = 18 \end{cases}$

D) $\begin{cases} 4w + 5t = 41 \\ w + t = 18 \end{cases}$

An investor receives dividend payments that are 5% of a business's monthly profits. The table below shows the amount D, in dollars, the investor received and the business profits P, in dollars, for each of four months:

Monthly Profits and Dividend Payments

Month	P, in dollars	D, in dollars
January	100,000	5,000
February	120,000	6,000
March	144,000	7,200
April	172,800	8,640

Which of the following equations represents the relationship between P and D?

A) $D = 50P$

B) $D = 5P$

C) $D = 0.5P$

D) $D = 0.05P$

When $f(x) = \sqrt{x}$, $g(x) = x^2$, what is the value of $f[g(-25)]$?

A) 25

B) –25

C) $\sqrt{25}$

D) Not a real number

	Mozart	Haydn
Symphony	61	106
Concerto	79	45
Piano music	68	19
Opera	23	29

The table above shows the number of each of four different types of music composed by Mozart and Haydn. If a concerto is selected at random, what is the probability that Mozart was not the composer?

A) $\dfrac{199}{430}$

B) $\dfrac{45}{430}$

C) $\dfrac{45}{199}$

D) $\dfrac{45}{124}$

This week, Jenny can work a maximum of 36 hours and need to earn at least $450. Her job in a mall pays $12 per hour, and her job in a college pays $15 per hour. If x represents the number of hours worked in the mall and y represents the number of hours worked in the college, which of the following system of inequalities represents the situation?

A) $\begin{cases} x + y \leq 36 \\ x + y \geq 450 \end{cases}$

B) $\begin{cases} x + y \geq 36 \\ 12x + 15y \leq 450 \end{cases}$

C) $\begin{cases} x + y \leq 36 \\ 15x + 12y \geq 450 \end{cases}$

D) $\begin{cases} x + y \leq 36 \\ 12x + 15y \geq 450 \end{cases}$

Which of the following is equivalent to the expression $1 + x^2 - x^3 - x$?

A) $(x + 1)(x - 1)^2$

B) $(x - 1)(x + 1)^2$

C) $(1 - x)(x + 1)^2$

D) $(1 - x)(x^2 + 1)$

Questions 14–16 refer to the following information.

	1	2	3	4	5	total
Adults	13	35	103	82	17	250
Children	6	22	27	70	25	150
Total	19	57	130	152	42	400

On the first night of each movie's release, the manager of a movie theater asks the people who saw the movie to rate it on a scale of 1 (worst) to 5 (best). The table above summarizes the responses of all 400 viewers of one particular movie.

How does the median rating of the adults who saw the movie compare to the median rating of the children who saw the movie?

A) The median ratings are the same.

B) The median rating of the adults is greater.

C) The median rating of the children is greater.

D) The table does not give enough information to compare the medians.

15

What fraction of the adults surveyed gave a rating of no less than 4 to the movie?

A) $\dfrac{82}{250}$

B) $\dfrac{99}{250}$

C) $\dfrac{194}{400}$

D) $\dfrac{99}{194}$

16

If the theater manager assumes the surveyed group is representative of the first 8,000 people to view the movie in this theater, about how many of the 8,000 people would the theater manager expect to rate the movie a 2 or lower?

A) 380

B) 760

C) 1520

D) 3240

17

If a is a constant and $a < 0$, how many solutions does the equation $(x + a)^{\frac{1}{2}} = a$ have?

A) No solution

B) 1 distinct solution

C) 2 distinct solutions

D) Infinitely many solutions

Questions 18–19 refer to the following information.

Men: $t_{man} = -0.1569x + 361.8$

Women: $t_{woman} = -0.2514x + 555.6$

Since 1912, both men and women have competed in bicycle races at the Summer Olympic Games. The winning times for men and women from 1912 through 2012 can be modeled by the equations above, where t represents the winning time, in seconds, in year x.

18

According to the predictions from the equations, if the Summer Olympic Games had occurred in the year 2006, which of the following is closest to the number of seconds by which the men's winning time was less than the women's winning time?

A) 2

B) 3

C) 4

D) 5

151

According to the equations, which of the following conclusions can be drawn concerning the winning times for men and women from 1912 through 2012?

A) t_{man} decreases faster per year than t_{woman}, and the value $t_{woman} - t_{man}$ is increasing as x increases.

B) t_{man} decreases faster per year than t_{woman}, and the value $t_{woman} - t_{man}$ is decreasing as x increases.

C) t_{woman} decreases faster per year than t_{man}, and the value $t_{woman} - t_{man}$ is increasing as x increases.

D) t_{woman} decreases faster per year than t_{man}, and the value $t_{woman} - t_{man}$ is decreasing as x increases.

A manufacturer packages vegetables in cans that are in the shape of right cylinders with height of 15 centimeters and volume of 750 cubic centimeters. If the manufacturer reduces the volume of the cans to 600 cubic centimeters but keeps the area of the base the same, by how many centimeters do the height of the can decrease?

A) 3

B) 4

C) 5

D) 6

Bonny sets guidelines in his writings for the relationship between the height (rise) and tread (run) of stairs in a building. He wrote that the rise should be between 9 and 10 inches, inclusive, per step, and the run should be between 18 and 24 inches, inclusive. Which of the following could be the ratio of rise to run for a set of stairs that follows this guideline?

A) $\frac{1}{3}$

B) $\frac{3}{5}$

C) $\frac{1}{2}$

D) $\frac{27}{40}$

$$y = (x-1)(x-3)(x+5)$$

In the xy-plane, which of the following could be a graph of the equation above?

A)

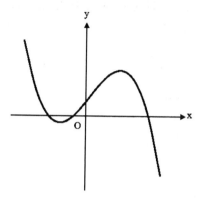

B)

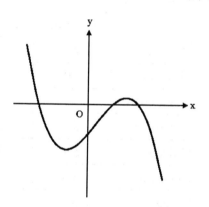

C)

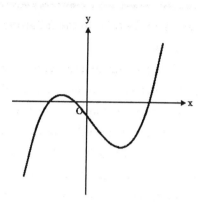

D)

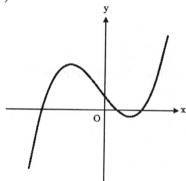

The payload package of a rocket is 5 feet above the ground before the rocket is launched from a space station. After launch, the rocket rises for 30 minutes. The height of the payload package is estimated to increase by 10 feet every 15 seconds. The function h gives the height $h(x)$, in feet, of the payload package above the ground in terms of the time x, in seconds, after the rocket is launched.

23

Which of the following represents $h(x)$?

A) $h(x)=x+5$

B) $h(x)=\dfrac{2}{3}x$

C) $h(x)=5+\dfrac{2}{3}x$

D) $h(x)=\dfrac{5+2x}{3}$

24

Which of the following is the best interpretation of the slope of the graph h?

A) The ratio of the height of the payload package, in feet above the ground, to the distance from the weather station, in feet

B) The increase of height of the payload package, in feet above the ground, each second

C) The height of weather station, in feet above the ground

D) The angle that the path of the balloon makes with the ground

25

$$2x - 3y = 11$$
$$3x + 5y = 23$$

The solution to the system of equations above is (x, y), what is the value of y?

A) 12

B) $\dfrac{13}{19}$

C) $\dfrac{5}{6}$

D) $\dfrac{3}{19}$

26

Day	Low tide	High tide
Monday	7.9	30.2
Tuesday	7.9	29.5
Wednesday	7.2	30.1
Thursday	6.6	30.8
Friday	5.9	31.5
Saturday	5.6	33.1
Sunday	6.9	31.5

According to the table, what is the difference between the mode of predicted high tide and the mode of predicted low tide over the 7-day period?

A) 23.6 feet

B) 23.9 feet

C) 24.2 feet

D) 24.9 feet

$$x^2 + y = 4x - 3$$
$$y = 1 - 2x$$

If (x_1, y_1) and (x_2, y_2) are two distinct solutions of the system of equations above, what is the value of $x_1 + x_2$?

A) 6

B) 4

C) –6

D) –4

In a scale drawing of a rectangular kitchen floor, the width of the floor is 2 inches and the length is 8 inches. If the width of the actual kitchen floor is w feet, which of the following function A could represent the area, in square feet, of the actual living room floor?

A) $A(w) = \dfrac{w^2}{8}$

B) $A(w) = 2w^2$

C) $A(w) = 4w^2$

D) $A(w) = 8w^2$

$$9x^2 + bx + 49 = 0$$

In the equation above, b is a constant. If the equation has no real solution, which of the following could be the value of b?

A) 7

B) 219

C) 42

D) –42

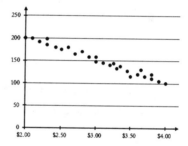

In the scatterplot above, each point represents the quantity n of a product sold by a business at price p, in dollars, for one of 30 days. Which of the following equations best models the relationship between price and quantity sold?

A) $n = 50p + 200$

B) $n = -50p + 200$

C) $n = -50p + 300$

D) $p = \dfrac{n}{50} + 2$

155

Four animals are used as models and carved into a granite mountain. The ratio of a sculpture's head length to the actual animal's head length is 40 feet to 6 inches. What is the sculpture's head length, in feet, for an animal with a head length of 9 inches?

Tony is buying prizes for a debate competition. He will buy one prize that costs $60 and four prizes that cost $15 each. The remainder of the prizes will cost $10 each. If Tony can spend no more than $300 on all the prizes, what is the largest total number of prizes he can buy for the competition? (Assume there is no sales tax.)

The graph of a linear function f in the xy-plane has a y-intercept of 9 and a slope of $-\frac{3}{4}$. What is the x-intercept of the graph of f?

Sample	1	2	3	4	5	6	7
Lead (ppm)	1523	847	1267	692	1401	1088	?

To determine whether to buy a plot of land, a farmer is having the soil tested for lead. The table above shows the amount of lead, in parts per million (ppm), in six of the seven samples of soil selected at random. If the average (arithmetic mean) amount of lead in the seven samples, in ppm, is no more than 1100, the farmer will buy the plot of land; otherwise, the farmer will not buy it. What is the greatest possible amount of lead, in ppm, that can be in the seventh sample if the farmer buys the land?

35

What is the radius of the circle in the *xy*-plane that has the center as (1, 5) and contains the point (5, 8)?

36

Component	Percentage (by weight)
Chromium	18%
Nickel	10%
Steel (98% iron, 2% carbon by weight)	72%

This alloy is *p* percent iron by weight. What is the value of *p*, to the nearest whole number?

Questions 37–38 refer to the following information.

A researcher conducted a study of the effects of taking a specific vitamin once a day on the blood pressure of elderly people in the state of Vermont. The sample used for the study include all residents at the Chestnut Hills Nursing Home, located in Vermont. The sample is made up of 38 male participants and 56 female participants. The distribution of the ages of all participants in the study is shown in the table below.

Age range	Number of participants
70–74	12
75–79	37
80–84	37
85 and older	8
Total	94

37

If the ratio of female to male participants is 4 to 1 for participants who are 80 years or older, how many participants 80 years or older are female?

38

The participants younger than 80 years old are *x* percent of all participants. What is the value of *x*, to the nearest whole number?

Answer Key

1. Equation

LEVEL 1

1) B	2) B	3) C	4) A	5) C	6) B	7) C	8) B
9) D	10) B	11) C	12) A	13) A	14) B	15) C	16) A
17) C	18) B	19) 10/3	20) B	21) 4	22) 60	23) A	24) C
25) D	26) D	27) B	28) D	29) 4	30) D	31) A	32) 2

LEVEL 2

1) D	2) C	3) B	4) C	5) A	6) 2	7) 237.5	8) B
9) B	10) 7	11) 1	12) 3.5	13) D	14) D	15) C	16) A
17) 181	18) 21	19) A	20) 7	21) A	22) 12	23) D	24) A
25) 9							

LEVEL 3

1) B	2) D	3) 150	4) C	5) C

2. Absolute Value

LEVEL 2

1) B	2) 4

3. Inequality

LEVEL 1

1) D	2) B	3) 3 or 5	4) 4 or 5 or 8	5) 5	6) D

LEVEL 2

1) D	2) C	3) 77 or 78 or 79 or 80	4) D	5) A
6) B	7) A	8) D	9) C	10) C

LEVEL 3

1) B	2) B

4. Linear Function

LEVEL 1

1) A	2) B	3) B	4) D	5) A	6) 14	7) A	8) B
9) B	10) A	11) B	12) C	13) 44	14) A	15) A	16) A
17) C	18) C	19) D	20) D	21) A	22) D	23) A	24) 9
25) C	26) C	27) C					

LEVEL 2

1) C	2) C	3) B	4) D	5) B	6) D	7) 50	8) B
9) C	10) B	11) D	12) B	13) C	14) A	15) B	16) B
17) A	18) D	19) A	20) D	21) C	22) D	23) A	24) C
25) B	26) 2	27) A	28) C	29) B	30) A	31) C	32) D
33) 4.08	34) B						

LEVEL 3

1) D	2) 3	3) A	4) B	5) A	6) 4/5	7) D	8) 1165
9) A	10) B	11) A	12) A	13) D	14) 12	15) C	16) B
17) A							

5. Exponential Function

LEVEL 2

1) A	2) A	3) C	4) B	5) D

LEVEL 3

1) C	2) C

6. Quadratic Function

LEVEL 1

1) 7	2) A	3) D	4) C	5) B

LEVEL 2

1) C	2) C	3) C	4) B	5) A	6) A	7) D	8) C
9) D	10) A	11) C	12) 13 or 5				

LEVEL 3
1) 2   2) A   3) B   4) A   5) 6   6) 2   7) 20   8) B
9) C   10) A   11) D

7. Function

LEVEL 1
1) 1   2) D   3) A   4) 10

LEVEL 2
1) A   2) 4 or 5 or 8   3) 0 or 1   4) 60   5) 1   6) C   7) D
8) B   9) B   10) A   11) B   12) 9   13) B

LEVEL 3
1) A   2) 8   3) 3/2   4) 3   5) A

8. Polynomial Factors and Graphs

LEVEL 2
1) C   2) C   3) A

LEVEL 3
1) C   2) A

9. Unit

LEVEL 1
1) A   2) B   3) D   4) D   5) B

LEVEL 2
1) D   2) C   3) B   4) 342

LEVEL 3
1) 1403   2) A

10. Percent

LEVEL 1
1) C   2) A   3) B

LEVEL 2
1) B   2) 3150   3) B   4) C   5) C

LEVEL 3
1) B

11. Proportion/Probability/Fraction

LEVEL 1
1) B   2) 2640   3) 0.65

LEVEL 2
1) C   2) A   3) C   4) B   5) C   6) A   7) 39.5   8) 19.9

LEVEL 3
1) D   2) A   3) C   4) B   5) A   6) C   7) D   8) D
9) B

12. Scatterplot

LEVEL 2
1) B   2) C   3) A   4) D   5) A   6) C   7) C   8) A
9) B   10) A   11) C   12) B   13) B   14) B   15) B   16) B

LEVEL 3
1) B   2) B   3) C   4) D   5) A

13. Box Plot

LEVEL 1
1) D

LEVEL 2
1) B

LEVEL 3
1) C   2) B

14. Statistics

LEVEL 1
1) D   2) D   3) 3744   4) C

LEVEL 2
1) D   2) D   3) B   4) C   5) D   6) B

LEVEL 3
1) C   2) C   3) C   4) C   5) A   6) B   7) B

15. Sample Survey

LEVEL 1
1) A

LEVEL 2
1) C   2) C   3) A   4) C

LEVEL 3
1) A   2) B   3) A

16. Polynomials

LEVEL 1
1) C   2) 4   3) D   4) A   5) C

LEVEL 2
1) C

LEVEL 3
1) A   2) B

17. Radicals and Rational Exponents

LEVEL 1
1) C   2) C   3) C

LEVEL 2
1) B   2) D   3) C   4) 2   5) A   6) C   7) A   8) C

LEVEL 3
1) D   2) C   3) B   4) A

18. Plane Geometry

LEVEL 1
1) B   2) A   3) A   4) A   5) D   6) 26 or 39 or 52

LEVEL 2
1) 40   2) B   3) B   4) B   5) A   6) C   7) D   8) B
9) D

LEVEL 3
1) C   2) C   3) 1.57   4) 4   5) A   6) B   7) A   8) A

19. Solid Geometry

LEVEL 2
1) D   2) A   3) 10

LEVEL 3
1) 6

20. Coordinate Geometry

LEVEL 1
1) C   2) C

LEVEL 2
1) C   2) D   3) C

21. Trigonometry

LEVEL 1
1) B

LEVEL 2
1) C

LEVEL 3
1) 3/4

22. Degree and Radian

LEVEL 1
1) 144

LEVEL 2
1) 60   2) 330

23. Complex Number

LEVEL 1
1) 8

LEVEL 3
1) 0

24. Others

LEVEL 2
1) B   2) 8.7   3) 0.3   4) A

Final Exam 1

Section 3 (No calculator)

1) B	2) C	3) C	4) D	5) B	6) B	7) D	8) B
9) C	10) B	11) B	12) D	13) B	14) D	15) C	16) 160
17) 32	18) 25	19) 2.5	20) 42				

Section 4 (Calculator)

1) B	2) A	3) A	4) B	5) C	6) B	7) D	8) D
9) C	10) B	11) B	12) A	13) B	14) B	15) C	16) A
17) B	18) D	19) C	20) B	21) C	22) A	23) B	24) A
25) B	26) D	27) C	28) B	29) C	30) C	31) 40	32) 24
33) 85	34) 13125	35) 6	36) 5142	37) 11.9	38) 22		

Final Exam 2

Section 3 (No calculator)

1) C	2) B	3) B	4) A	5) B	6) C	7) D	8) B
9) B	10) A	11) B	12) A	13) D	14) D	15) B	16) 3
17) 4	18) 0	19) 25	20) 2.1				

Section 4 (Calculator)

1) A	2) B	3) A	4) D	5) C	6) B	7) A	8) A
9) D	10) A	11) D	12) D	13) D	14) C	15) B	16) C
17) A	18) C	19) D	20) A	21) C	22) D	23) C	24) B
25) B	26) A	27) A	28) C	29) A	30) C	31) 60	32) 23
33) 12	34) 882	35) 5	36) 71	37) 36	38) 52		